Physics of Continuous Media

A collection of problems with solutions for physics students

Physics of Continuous Media

A collection of problems with solutions for physics students

G E Vekstein

Institute of Nuclear Physics,
USSR Academy of Sciences, Novosibirsk, USSR

CRC Press
Taylor & Francis Group
Boca Raton London New York

CRC Press is an imprint of the
Taylor & Francis Group, an **informa** business

First published 1992 by IOP Publishing Ltd

Published 2019 by CRC Press
Taylor & Francis Group
6000 Broken Sound Parkway NW, Suite 300
Boca Raton, FL 33487-2742

First issued in paperback 2019

ISBN 13: 978-0-367-45049-6 (pbk)
ISBN 13: 978-0-7503-0140-4 (hbk)

British Library Cataloguing in Publication Data
Vekstein, G.E.
Physics of continuous media: A collection of problems with solutions for physics students.
I. Title
532

ISBN 0-7503-0140-6
ISBN 0-7503-0141-4 pbk

Library of Congress Cataloging-in-Publication Data are available

Typeset by P&R Typesetters Ltd, Salisbury, Wiltshire

Contents

Preface

This textbook, aimed at undergraduate and postgraduate students in physics and applied mathematics, is based on lectures and tutorials given for several years at the Physics Department of Novosibirsk State University.

It is constructed as a set of problems followed by detailed solutions and may act as a complementary text for standard courses on the physics of continuous media. This form permits the author to present new information without repetition of what is available in the existing literature. For instance, it contains some model problems illucidating the physical nature of phenomena traditionally treated only phenomenologically (such as the energy and momentum of electromagnetic waves in a medium, waves with negative energy, natural optical activity, etc). Most of the problems included are original; some are 'standard' problems useful for beginners, which are developed further; others originated didactically from physical effects considered in the current literature (e.g., Landau damping of sound waves in a fluid with gas bubbles, Fredericksz transition in a nematic liquid crystal, resonant excitation of surface waves by the wind, non-linear interaction of waves, and so on).

Each chapter begins with a minimum of definitions and equations, since all the required information is available in existing comprehensive textbooks (several of which are mentioned there). Some more particular references are given through the text.

I am extremely grateful to many of my colleagues in Novosibirsk (D Ryutov, G Stupakov, B Breizman, I Khriplovich) for their critical comments and constructive suggestions, as well as to E Priest from the University of St Andrews, Scotland, who encouraged me to prepare this English version of the book.

G E Vekstein
Novosibirsk, September 1990

PART I

Electrodynamics of Continuous Media

Chapter 1

Vectors, Tensors, Fourier Transformation

Recommended textbook:
Spiegel M R 1974 *Vector Analysis* (*Schaum Outline Series*) (New York: McGraw-Hill)

Vector Operations in Cylindrical Polar Coordinates r, φ, z

$$\operatorname{grad} f \equiv \nabla f = \boldsymbol{e}_r \frac{\partial f}{\partial r} + \boldsymbol{e}_\varphi \frac{1}{r}\frac{\partial f}{\partial \varphi} + \boldsymbol{e}_z \frac{\partial f}{\partial z}$$

$$\operatorname{div} \boldsymbol{A} \equiv (\nabla \cdot \boldsymbol{A}) = \frac{1}{r}\frac{\partial}{\partial r}(rA_r) + \frac{1}{r}\frac{\partial A_\varphi}{\partial \varphi} + \frac{\partial A_z}{\partial z}$$

Laplacian

$$\nabla^2 f = \operatorname{div}(\nabla f) = \frac{1}{r}\frac{\partial}{\partial r}\left(r\frac{\partial f}{\partial r}\right) + \frac{1}{r^2}\frac{\partial^2 f}{\partial \varphi^2} + \frac{\partial^2 f}{\partial z^2}$$

$$(\operatorname{curl} \boldsymbol{A})_r = (\nabla \times \boldsymbol{A})_r = \frac{1}{r}\frac{\partial A_z}{\partial \varphi} - \frac{\partial A_\varphi}{\partial z}$$

$$(\operatorname{curl} \boldsymbol{A})_\varphi = (\nabla \times \boldsymbol{A})_\varphi = \frac{\partial A_r}{\partial z} - \frac{\partial A_z}{\partial r}$$

$$(\operatorname{curl} \boldsymbol{A})_z = (\nabla \times \boldsymbol{A})_z = \frac{1}{r}\frac{\partial}{\partial r}(rA_\varphi) - \frac{1}{r}\frac{\partial A_r}{\partial \varphi} \tag{1.1}$$

$$(\nabla^2 \boldsymbol{A})_r = \nabla^2 A_r - \frac{A_r}{r^2} - \frac{2}{r^2}\frac{\partial A_\varphi}{\partial \varphi}$$

$$(\nabla^2 \boldsymbol{A})_\varphi = \nabla^2 A_\varphi - \frac{A_\varphi}{r^2} + \frac{2}{r^2}\frac{\partial A_r}{\partial \varphi}$$

$$(\nabla^2 \boldsymbol{A})_z = \nabla^2 A_z$$

Vector Operations in Spherical Polar Coordinates r, θ and φ

$$\operatorname{grad} f = \nabla f = \boldsymbol{e}_r \frac{\partial f}{\partial r} + \boldsymbol{e}_\theta \frac{1}{r} \frac{\partial f}{\partial \theta} + \boldsymbol{e}_\varphi \frac{1}{r \sin\theta} \frac{\partial f}{\partial \varphi}$$

$$\operatorname{div} \boldsymbol{A} = (\nabla \cdot \boldsymbol{A}) = \frac{1}{r^2} \frac{\partial}{\partial r}(r^2 A_r) + \frac{1}{r \sin\theta} \frac{\partial}{\partial \theta}[(\sin\theta) A_\theta] + \frac{1}{r \sin\theta} \frac{\partial A_\varphi}{\partial \varphi}$$

Laplacian

$$\nabla^2 f = \frac{1}{r^2} \frac{\partial}{\partial r}\left(r^2 \frac{\partial f}{\partial r}\right) + \frac{1}{r^2 \sin\theta} \frac{\partial}{\partial \theta}\left(\sin\theta \frac{\partial f}{\partial \theta}\right) + \frac{1}{r^2 \sin^2\theta} \frac{\partial^2 f}{\partial \varphi^2}$$

$$(\operatorname{curl} \boldsymbol{A})_r = (\nabla \times \boldsymbol{A})_r = \frac{1}{r \sin\theta}\left(\frac{\partial}{\partial \theta}[(\sin\theta) A_\varphi] - \frac{\partial A_\theta}{\partial \varphi}\right)$$

$$(\operatorname{curl} \boldsymbol{A})_\theta = (\nabla \times \boldsymbol{A})_\theta = \frac{1}{r \sin\theta} \frac{\partial A_r}{\partial \varphi} - \frac{1}{r} \frac{\partial}{\partial r}(r A_\varphi)$$

$$(\operatorname{curl} \boldsymbol{A})_\varphi = (\nabla \times \boldsymbol{A})_\varphi = \frac{1}{r}\left(\frac{\partial}{\partial r}(r A_\theta) - \frac{\partial A_r}{\partial \theta}\right) \tag{1.2}$$

$$(\nabla^2 \boldsymbol{A})_r = \nabla^2 A_r - \frac{2}{r^2}\left(A_r + \frac{1}{\sin\theta} \frac{\partial}{\partial \theta}[(\sin\theta) A_\theta] + \frac{1}{\sin\theta} \frac{\partial A_\varphi}{\partial \varphi}\right)$$

$$(\nabla^2 \boldsymbol{A})_\theta = \nabla^2 A_\theta + \frac{2}{r^2}\left(\frac{\partial A_r}{\partial \theta} - \frac{A_\theta}{2 \sin^2\theta} - \frac{\cos\theta}{\sin^2\theta} \frac{\partial A_\varphi}{\partial \varphi}\right)$$

$$(\nabla^2 \boldsymbol{A})_\varphi = \nabla^2 A_\varphi + \frac{2}{r^2 \sin\theta}\left(\frac{\partial A_r}{\partial \varphi} + (\cot\theta) \frac{\partial A_\theta}{\partial \varphi} - \frac{A_\varphi}{2 \sin\theta}\right)$$

Some Identities of a Vector Analysis

$$\mathrm{div}(f\boldsymbol{A}) = f\,\mathrm{div}\,\boldsymbol{A} + \boldsymbol{A}\cdot\mathrm{grad}\,f \tag{1.3a}$$

$$\mathrm{curl}(f\boldsymbol{A}) = f\,\mathrm{curl}\,\boldsymbol{A} - \boldsymbol{A}\times\nabla f \tag{1.3b}$$

$$\mathrm{div}(\boldsymbol{A}\times\boldsymbol{B}) = \boldsymbol{B}\,\mathrm{curl}\,\boldsymbol{A} - \boldsymbol{A}\,\mathrm{curl}\,\boldsymbol{B} \tag{1.3c}$$

$$\mathrm{curl}(\boldsymbol{A}\times\boldsymbol{B}) = \boldsymbol{A}\,\mathrm{div}\,\boldsymbol{B} - \boldsymbol{B}\,\mathrm{div}\,\boldsymbol{A} + (\boldsymbol{B}\cdot\nabla)\boldsymbol{A} - (\boldsymbol{A}\cdot\nabla)\boldsymbol{B} \tag{1.3d}$$

$$\mathrm{grad}(\boldsymbol{A}\cdot\boldsymbol{B}) = \boldsymbol{A}\times\mathrm{curl}\,\boldsymbol{B} + \boldsymbol{B}\times\mathrm{curl}\,\boldsymbol{A} + (\boldsymbol{B}\cdot\nabla)\boldsymbol{A} + (\boldsymbol{A}\cdot\nabla)\boldsymbol{B} \tag{1.3e}$$

Hermitian and AntiHermitian Tensors

A tensor of rank two T_{ik} is a Hermitian tensor if $T_{ki} = T_{ik}^*$, and is an antiHermitian tensor if $T_{ki} = -T_{ik}^*$. Any tensor T_{ik} may be represented as a sum of a Hermitian tensor $T_{ik}^{(H)}$ and an antiHermitian tensor $T_{ik}^{(A)}$ in the following manner:

$$T_{ik} = T_{ik}^{(H)} + T_{ik}^{(A)} \tag{1.4}$$

$$T_{ik}^{(H)} = \tfrac{1}{2}(T_{ik} + T_{ki}^*) \qquad T_{ik}^{(A)} = \tfrac{1}{2}(T_{ik} - T_{ki}^*).$$

The vector $\boldsymbol{A}$ is called an *eigenvector* of a matrix operator T_{ik}, if $T_{ik}A_k = \lambda A_i$. In this case a number λ is an *eigenvalue* of the matrix T_{ik}.

If a matrix is Hermitian all its eigenvalues are real, and vice versa, all eigenvalues of an antiHermitian matrix are pure imaginary.

Problem 1

Calculate the averaged values for the following expressions: $(\boldsymbol{a}\cdot\boldsymbol{n})\cdot\boldsymbol{n}$, $(\boldsymbol{a}\times\boldsymbol{n})^2$ and $(\boldsymbol{a}\times\boldsymbol{n})\cdot(\boldsymbol{b}\cdot\boldsymbol{n})$, if $\boldsymbol{a}$ and $\boldsymbol{b}$ are the given vectors while $\boldsymbol{n}$ is a unit vector with a random but isotropically distributed space orientation.

It is convenient to use tensor notations here. For instance,

$$\langle(\boldsymbol{a}\cdot\boldsymbol{n})\boldsymbol{n}\rangle = \langle a_k n_k n_i\rangle = a_k\langle n_k n_i\rangle$$

where the sign $\langle\ \rangle$ means the average over all possible orientations of the vector $\boldsymbol{n}$. Because its orientations are distributed isotropically, the tensor of rank two $\langle n_k n_i\rangle$ is the same in any coordinate system, that is it is an invariant tensor. Therefore, $\langle n_k n_i\rangle$ is proportional to the unit tensor of rank two δ_{ik}: $\langle n_k n_i\rangle = A\delta_{ik}$. The constant A may be determined by considering the traces of these tensors. Since $\boldsymbol{n}$ is a unit vector, its trace $n_i n_i = 1$. Thus,

$A\delta_{ii} = 3A = 1$, and $A = \frac{1}{3}$. The final result is $a_k\langle n_k n_i\rangle = \frac{1}{3}a_i$, or, in vector notation, $\langle(\boldsymbol{a}\cdot\boldsymbol{n})\boldsymbol{n}\rangle = \frac{1}{3}\boldsymbol{a}$.

All other expressions may be averaged in the same way:

$$\langle(\boldsymbol{a}\times\boldsymbol{n})^2\rangle = \langle e_{ikl}a_k n_l e_{ipq}a_p n_q\rangle = a_k a_p e_{ikl}e_{ipq}\langle n_l n_q\rangle = \tfrac{1}{3}\delta_{lq}a_k a_p e_{ikl}e_{ipq}$$
$$= \tfrac{1}{3}a_k a_p e_{ikl}e_{ipl}$$

where e_{ipl} is the unit antisymmetric pseudotensor of rank three. Taking into account that

$$e_{ikl}e_{ipl} = 2\delta_{kp}$$

we obtain that $\langle(\boldsymbol{a}\times\boldsymbol{n})^2\rangle = \frac{2}{3}a_k a_p\delta_{kp} = \frac{2}{3}a^2$.

$$\langle(\boldsymbol{a}\times\boldsymbol{n})\cdot(\boldsymbol{b}\cdot\boldsymbol{n})\rangle = \langle e_{ikl}a_k n_l b_m n_m\rangle = e_{ikl}a_k b_m\langle n_l n_m\rangle = e_{ikl}a_k b_m\tfrac{1}{3}\delta_{lm}$$
$$= \tfrac{1}{3}e_{ikl}a_k b_l = \tfrac{1}{3}(\boldsymbol{a}\times\boldsymbol{b}).$$

Problem 2

Find the values, averaged over time, for the tensors $E_\alpha E_\beta$, $B_\alpha B_\beta$, $E_\alpha B_\beta$, where the vectors $\boldsymbol{E}$ and $\boldsymbol{B}$ are the electric field and the magnetic induction of a circularly polarized electromagnetic wave, propagating in a vacuum with a wave vector $\boldsymbol{k}$. The amplitude of the electric field is equal to E_0.

Let us start with the particular coordinate system in which the vector $\boldsymbol{k}$ is directed along the z-axis. In this case the vectors $\boldsymbol{E}$ and $\boldsymbol{B}$ have only $(x - y)$ components with

$$E_x = E_0\cos\omega t \qquad E_y = E_0\sin\omega t$$
$$B_x = -E_0\sin\omega t \qquad B_y = E_0\cos\omega t \qquad \omega = kc.$$

Thus in this coordinate system the time-averaged tensors $\langle E_\alpha E_\beta\rangle$, $\langle B_\alpha B_\beta\rangle$ and $\langle E_\alpha B_\beta\rangle$ take the following form:

$$\langle E_\alpha E_\beta\rangle = \langle B_\alpha B_\beta\rangle = \begin{pmatrix} E_0^2/2 & 0 & 0 \\ 0 & E_0^2/2 & 0 \\ 0 & 0 & 0 \end{pmatrix}$$

and

$$\langle E_\alpha B_\beta\rangle = \begin{pmatrix} 0 & E_0^2/2 & 0 \\ -E_0^2/2 & 0 & 0 \\ 0 & 0 & 0 \end{pmatrix}.$$

Now it is possible to write these expressions in the tensor-invariant notation:

$$\langle E_\alpha E_\beta \rangle = \langle B_\alpha B_\beta \rangle = \frac{E_0^2}{2}(\delta_{\alpha\beta} - n_\alpha n_\beta)$$

$$\langle E_\alpha B_\beta \rangle = \frac{E_0^2}{2} e_{\alpha\beta\gamma} n_\gamma$$

where $\boldsymbol{n} = \boldsymbol{k}/k$ is the unit vector in the direction in which the wave propagates. It is worth noting that the last combination is a pseudotensor since $\boldsymbol{B}$ is an axial vector.

Problem 3

Determine the electromagnetic field of a point charge q moving with a constant velocity $\boldsymbol{v}$ in a vacuum by solving the Maxwell equations with the help of a Fourier transformation.

Let us write the Maxwell equations

$$\text{curl } \boldsymbol{E} = -\frac{1}{c}\frac{\partial \boldsymbol{B}}{\partial t} \qquad \text{curl } \boldsymbol{B} = \frac{1}{c}\frac{\partial \boldsymbol{E}}{\partial t} + \frac{4\pi}{c}\boldsymbol{j} \tag{1.5}$$

with $\boldsymbol{j} = q\boldsymbol{v}\cdot\delta(\boldsymbol{r} - \boldsymbol{v}t)$, and make the Fourier transformation from the variables $(\boldsymbol{r}, t)$ to the variables $(\boldsymbol{k}, \omega)$ by the following rule:

$$\begin{aligned} \boldsymbol{E}(\boldsymbol{r}, t) &= \frac{1}{(2\pi)^2}\int \boldsymbol{E}(\boldsymbol{k}, \omega)\, \mathrm{e}^{\mathrm{i}(\boldsymbol{k}\cdot\boldsymbol{r} - \omega t)}\, \mathrm{d}\boldsymbol{k}\, \mathrm{d}\omega \\ \boldsymbol{B}(\boldsymbol{r}, t) &= \frac{1}{(2\pi)^2}\int \boldsymbol{B}(\boldsymbol{k}, \omega)\, \mathrm{e}^{\mathrm{i}(\boldsymbol{k}\cdot\boldsymbol{r} - \omega t)}\, \mathrm{d}\boldsymbol{k}\, \mathrm{d}\omega. \end{aligned} \tag{1.6}$$

The inverse transformation takes the form

$$\begin{aligned} \boldsymbol{E}(\boldsymbol{k}, \omega) &= \frac{1}{(2\pi)^2}\int \boldsymbol{E}(\boldsymbol{r}, t)\, \mathrm{e}^{-\mathrm{i}(\boldsymbol{k}\cdot\boldsymbol{r} - \omega t)}\, \mathrm{d}\boldsymbol{r}\, \mathrm{d}t \\ \boldsymbol{B}(\boldsymbol{k}, \omega) &= \frac{1}{(2\pi)^2}\int \boldsymbol{B}(\boldsymbol{r}, t)\, \mathrm{e}^{-\mathrm{i}(\boldsymbol{k}\cdot\boldsymbol{r} - \omega t)}\, \mathrm{d}\boldsymbol{r}\, \mathrm{d}t. \end{aligned} \tag{1.7}$$

This procedure is valid for any function of $\boldsymbol{r}$ and t, that tends sufficiently rapidly to zero as $r \to \infty$, $t \to \pm\infty$, and will be used in what follows.

After the substitution of the $\boldsymbol{E}$ and $\boldsymbol{B}$ in the form of equation (1.6) into (1.5) and taking into account the identity (1.3b) we obtain the following relations for the Fourier transforms of $\boldsymbol{E}$ and $\boldsymbol{B}$:

$$\boldsymbol{B}(\boldsymbol{k}, \omega) = \frac{c}{\omega}[\boldsymbol{k} \times \boldsymbol{E}(\boldsymbol{k}, \omega)] \tag{1.8a}$$

$$[\boldsymbol{k}\times\boldsymbol{B}(\boldsymbol{k},\omega)] = -\frac{\omega}{c}\boldsymbol{E}(\boldsymbol{k},\omega) - \frac{4\pi\mathrm{i}}{c}\boldsymbol{j}(\boldsymbol{k},\omega) \tag{1.8b}$$

where, according to the rule (1.7), the Fourier transform of the current is

$$\boldsymbol{j}(\boldsymbol{k},\omega) = \frac{q\boldsymbol{v}}{(2\pi)^2}\int\delta(\boldsymbol{r}-\boldsymbol{v}t)\,\mathrm{e}^{-\mathrm{i}(\boldsymbol{k}\cdot\boldsymbol{r}-\omega t)}\,\mathrm{d}\boldsymbol{r}\,\mathrm{d}t$$

$$= \frac{q\boldsymbol{v}}{(2\pi)^2}\int\mathrm{e}^{\mathrm{i}(\omega-\boldsymbol{k}\cdot\boldsymbol{v})}\,\mathrm{d}t = \frac{q\boldsymbol{v}}{2\pi}\delta(\omega-\boldsymbol{k}\cdot\boldsymbol{v}).$$

Eliminating $\boldsymbol{B}$ from equation (1.8*b*) we have

$$\boldsymbol{k}(\boldsymbol{k}\cdot\boldsymbol{E}) - k^2\boldsymbol{E} = -\frac{\omega^2}{c^2}\boldsymbol{E} - \frac{4\pi\mathrm{i}\omega}{c^2}\boldsymbol{j}.$$

The value of $(\boldsymbol{k}\cdot\boldsymbol{E})$ may be determined from the scalar product of equation (1.8*b*) and $\boldsymbol{k}$, so

$$\boldsymbol{k}\cdot\boldsymbol{E} = -\frac{4\pi\mathrm{i}}{\omega}(\boldsymbol{k}\cdot\boldsymbol{j}).$$

Thus the Fourier transforms of $\boldsymbol{E}$ and $\boldsymbol{B}$ are as follows:

$$\boldsymbol{E}(\boldsymbol{k},\omega) = \frac{2\mathrm{i}q\delta(\omega-\boldsymbol{k}\cdot\boldsymbol{v})}{k^2-(\boldsymbol{k}\cdot\boldsymbol{v})^2/c^2}[-\boldsymbol{k}+\boldsymbol{v}(\boldsymbol{k}\cdot\boldsymbol{v})/c^2]$$

$$\boldsymbol{B}(\boldsymbol{k},\omega) = \frac{c}{\omega}[\boldsymbol{k}\times\boldsymbol{E}] = \frac{1}{c}[\boldsymbol{v}\times\boldsymbol{E}(\boldsymbol{k},\omega)].$$

Now we obtain from equation (1.6) that

$$\boldsymbol{E}(\boldsymbol{r},t) = \frac{2\mathrm{i}q}{(2\pi)^2}\int\mathrm{e}^{\mathrm{i}(\boldsymbol{k}\cdot\boldsymbol{r}-\omega t)}\,\mathrm{d}\boldsymbol{k}\,\mathrm{d}\omega\,\delta(\omega-\boldsymbol{k}\cdot\boldsymbol{v})\frac{[-\boldsymbol{k}+\boldsymbol{v}(\boldsymbol{k}\cdot\boldsymbol{v})/c^2]}{k^2-(\boldsymbol{k}\cdot\boldsymbol{v})^2/c^2}$$

$$= \frac{2\mathrm{i}q}{(2\pi)^2}\int\mathrm{e}^{\mathrm{i}[\boldsymbol{k}\cdot\boldsymbol{r}-(\boldsymbol{k}\cdot\boldsymbol{v})t]}\,\mathrm{d}\boldsymbol{k}\left(\frac{-\boldsymbol{k}+\boldsymbol{v}(\boldsymbol{k}\cdot\boldsymbol{v})/c^2}{k^2-(\boldsymbol{k}\cdot\boldsymbol{v})^2/c^2}\right). \tag{1.9}$$

To calculate the last integral let us note that in the particular case with $\boldsymbol{v}=0$ it must be equal to the simple Coulomb field, so that

$$-\frac{2\mathrm{i}q}{(2\pi)^2}\int\mathrm{e}^{\mathrm{i}\boldsymbol{k}\cdot\boldsymbol{r}}\frac{\boldsymbol{k}}{k^2}\,\mathrm{d}\boldsymbol{k} = \frac{q\boldsymbol{r}}{r^3}. \tag{1.10}$$

Let the charge move along the x-axis and introduce the vector $\boldsymbol{R} = \{(x-vt), y, z\}$. Then the integral in (1.9) may be written as

$$\int\mathrm{e}^{\mathrm{i}\boldsymbol{k}\cdot\boldsymbol{R}}\,\mathrm{d}\boldsymbol{k}\frac{[-\boldsymbol{k}+\boldsymbol{v}(\boldsymbol{k}\cdot\boldsymbol{v})/c^2]}{k^2-(\boldsymbol{k}\cdot\boldsymbol{v})^2/c^2} = -\int\frac{\mathrm{e}^{\mathrm{i}\boldsymbol{k}_*\cdot\boldsymbol{R}_*}}{k_*^2}\gamma\,\mathrm{d}\boldsymbol{k}_*\left\{\frac{k_{*x}}{\gamma}, \boldsymbol{k}_{*\perp}\right\}$$

where

$$\boldsymbol{R}_* = \{\gamma R_x, \boldsymbol{R}_\perp\} \qquad \boldsymbol{k}_* = \{k_x/\gamma, \boldsymbol{k}_\perp\}.$$

It is now seen by comparing this expression with (1.10) that

$$\boldsymbol{E}(\boldsymbol{r}, t) = \frac{\gamma q}{R_*^3}\left\{\frac{R_{*x}}{\gamma}, \boldsymbol{R}_{*\perp}\right\} = \frac{\gamma q \boldsymbol{R}}{[\gamma^2(x - vt)^2 + y^2 + z^2]^{3/2}}.$$

The magnetic induction is

$$\boldsymbol{B}(\boldsymbol{r}, t) = \frac{1}{c}[\boldsymbol{v} \times \boldsymbol{E}(\boldsymbol{r}, t)].$$

Chapter 2

The Tensor of Dielectric Permeability. Electromagnetic Waves in Media

Recommended textbooks:
Landau L D and Lifshitz E M 1984 *Course of Theoretical Physics*, vol 8, *Electrodynamics of Continuous Media* (Oxford: Pergamon)
Jackson J D 1975 *Classical Electrodynamics* (New York: Wiley)

In the linear electrodynamics of continuous media the current density $\boldsymbol{j}(\boldsymbol{r}, t)$ is proportional to the electric field strength $\boldsymbol{E}(\boldsymbol{r}, t)$: $\boldsymbol{j} = \hat{\sigma}\boldsymbol{E}$, where $\hat{\sigma}$ is a linear operator that in the most general case takes the following form:

$$j_\alpha(\boldsymbol{r}, t) = \int \sigma_{\alpha\beta}(\boldsymbol{r}, \boldsymbol{r}', t, t')E_\beta(\boldsymbol{r}', t')\,\mathrm{d}\boldsymbol{r}'\,\mathrm{d}t'. \tag{2.1}$$

By introducing the electric induction

$$\mathscr{D}(\boldsymbol{r}, t) \equiv \boldsymbol{E}(\boldsymbol{r}, t) + 4\pi \int_{-\infty}^{t} \boldsymbol{j}(\boldsymbol{r}, t')\,\mathrm{d}t' \tag{2.2}$$

the Maxwell equations may be written as follows:

$$\begin{aligned} \operatorname{curl} \boldsymbol{E} &= -\frac{1}{c}\frac{\partial \boldsymbol{B}}{\partial t} \\ \operatorname{curl} \boldsymbol{B} &= \frac{1}{c}\frac{\partial \boldsymbol{E}}{\partial t} + \frac{4\pi}{c}\boldsymbol{j} = \frac{1}{c}\frac{\partial \mathscr{D}}{\partial t} \equiv \frac{1}{c}\frac{\partial}{\partial t}(\hat{\varepsilon}\boldsymbol{E}). \end{aligned} \tag{2.3}$$

The dielectric permeability operator $\hat{\varepsilon}$ has a form analogous to (2.1):

$$\mathscr{D}_\alpha(\boldsymbol{r}, t) \equiv \hat{\varepsilon}\boldsymbol{E} = E_\alpha(\boldsymbol{r}, t) + \int \varepsilon_{\alpha\beta}(\boldsymbol{r}, \boldsymbol{r}', t, t')E_\beta(\boldsymbol{r}', t')\,\mathrm{d}\boldsymbol{r}'\,\mathrm{d}t'. \tag{2.4}$$

In the case of homogeneous and stationary (with properties independent of time) media the tensor functions $\sigma_{\alpha\beta}$ and $\varepsilon_{\alpha\beta}$ in equations (2.1) and (2.4) take the following form:

$$\sigma_{\alpha\beta}(\boldsymbol{r}, \boldsymbol{r}', t, t') = \sigma_{\alpha\beta}(\boldsymbol{\rho}, \tau)$$
$$\varepsilon_{\alpha\beta}(\boldsymbol{r}, \boldsymbol{r}', t, t') = \varepsilon_{\alpha\beta}(\boldsymbol{\rho}, \tau) \tag{2.5}$$
$$\boldsymbol{\rho} = \boldsymbol{r} - \boldsymbol{r}' \qquad \tau = t - t'.$$

It is convenient to use Fourier transforms in this case, since in $(\boldsymbol{k}, \omega)$ representation the operators $\hat{\sigma}$ and $\hat{\varepsilon}$ become simple multipliers: *the conductivity tensor* $\sigma_{\alpha\beta}(\boldsymbol{k}, \omega)$ and the *dielectric permeability tensor* $\varepsilon_{\alpha\beta}(\boldsymbol{k}, \omega)$:

$$j_\alpha(\boldsymbol{k}, \omega) = \sigma_{\alpha\beta}(\boldsymbol{k}, \omega) E_\beta(\boldsymbol{k}, \omega) \qquad \mathscr{D}_\alpha(\boldsymbol{k}, \omega) = \varepsilon_{\alpha\beta}(\boldsymbol{k}, \omega) E_\beta(\boldsymbol{k}, \omega). \tag{2.6}$$

The following relations for these tensors and functions $\sigma_{\alpha\beta}(\boldsymbol{\rho}, \tau)$ and $\varepsilon_{\alpha\beta}(\boldsymbol{\rho}, \tau)$ are valid:

$$\sigma_{\alpha\beta}(\boldsymbol{k}, \omega) = \int \sigma_{\alpha\beta}(\boldsymbol{\rho}, \tau)\, \mathrm{e}^{-\mathrm{i}(\boldsymbol{k}\cdot\boldsymbol{\rho} - \omega\tau)}\, \mathrm{d}\boldsymbol{\rho}\, \mathrm{d}\tau$$
$$\varepsilon_{\alpha\beta}(\boldsymbol{k}, \omega) = \delta_{\alpha\beta} + \int \varepsilon_{\alpha\beta}(\boldsymbol{\rho}, \tau)\, \mathrm{e}^{-\mathrm{i}(\boldsymbol{k}\cdot\boldsymbol{\rho} - \omega\tau)}\, \mathrm{d}\boldsymbol{\rho}\, \mathrm{d}\tau \tag{2.7}$$

and, according to the notation (2.2),

$$\varepsilon_{\alpha\beta}(\boldsymbol{k}, \omega) = \delta_{\alpha\beta} + \frac{4\pi\mathrm{i}}{\omega}\, \sigma_{\alpha\beta}(\boldsymbol{k}, \omega). \tag{2.8}$$

In terms of Fourier transforms the Maxwell equations (2.3) are equivalent to the following linear system:

$$L_{\alpha\beta} E_\beta \equiv \left(k_\alpha k_\beta - k^2 \delta_{\alpha\beta} + \frac{\omega^2}{c^2}\, \varepsilon_{\alpha\beta} \right) E_\beta = 0 \tag{2.9}$$

so the *dispersion equation* that determines a link between the wave vector $\boldsymbol{k}$ and the frequency ω of an electromagnetic wave in a medium takes the form

$$\det \| L_{\alpha\beta}(\boldsymbol{k}, \omega) \| = 0. \tag{2.10}$$

In the case of isotropic and inversion-invariant media (inversion invariant means that a medium is identical to its stereoisomeric form) the most general form of the dielectric permeability tensor is as follows:

$$\varepsilon_{\alpha\beta}(\boldsymbol{k}, \omega) = \varepsilon_\parallel(k, \omega) \frac{k_\alpha k_\beta}{k^2} + \varepsilon_\perp(k, \omega) \left(\delta_{\alpha\beta} - \frac{k_\alpha k_\beta}{k^2} \right). \tag{2.11}$$

The electromagnetic waves in such a medium may be separated into *longitudinal waves* (l) with $\boldsymbol{E} \| \boldsymbol{k}$ and *transverse waves* (t) with $\boldsymbol{E} \perp \boldsymbol{k}$. Their dispersion equations follow from (2.10) and (2.11):

$$\varepsilon_\parallel(k, \omega_\mathrm{l}) = 0 \qquad k^2 = \frac{\omega_\mathrm{t}^2}{c^2}\, \varepsilon_\perp(k, \omega_\mathrm{t}). \tag{2.12}$$

The damping of the electromagnetic wave in a medium is determined by the antiHermitian part (see equation (1.4)) of the tensor $\varepsilon_{\alpha\beta}$, so the dissipation power per unit volume is

$$Q = -\frac{i\omega}{8\pi}\varepsilon_{\alpha\beta}^{(A)}E_{0\alpha}^{*}E_{0\beta}. \tag{2.13}$$

In the so-called *transparency ranges*, where the damping is small ($\varepsilon_{\alpha\beta}^{(A)} \ll \varepsilon_{\alpha\beta}^{(H)}$), the energy per unit volume W and the momentum per unit volume $\boldsymbol{p}$ of the electromagnetic wave may be defined as:

$$\begin{aligned} W &= \frac{1}{16\pi\omega}\frac{\partial}{\partial\omega}[\omega^2\varepsilon_{\alpha\beta}^{(H)}(\boldsymbol{k},\omega)]E_{0\alpha}^{*}E_{0\beta} \\ \boldsymbol{p} &= \boldsymbol{k}W/\omega \end{aligned} \tag{2.14}$$

where $\varepsilon_{\alpha\beta}^{(H)}$ is the Hermitian part of the tensor $\varepsilon_{\alpha\beta}$, and $\boldsymbol{E}_0$ is the amplitude of the electric field in a wave.

Problem 4

Show that formula (2.1) is indeed the most general form of the link between the current and the electric field in linear electrodynamics.

Let $j_\alpha = \hat{\sigma}_{\alpha\beta}E_\beta$, where $\hat{\sigma}_{\alpha\beta}$ is an arbitrary linear operator upon space and time. With the help of Fourier transformation the electric field may be represented as

$$E_\beta(\xi) = \int A_\beta(\boldsymbol{q})\psi_{\boldsymbol{q}}(\xi)\,\mathrm{d}\boldsymbol{q} \qquad \xi = (\boldsymbol{r}, t) \qquad \boldsymbol{q} = (\boldsymbol{k}, \omega)$$

$$\psi_{\boldsymbol{q}}(\xi) = \frac{1}{(2\pi)^2}\mathrm{e}^{\mathrm{i}(\boldsymbol{k}\cdot\boldsymbol{r}-\omega t)}.$$

The coefficients are

$$A_\beta(\boldsymbol{q}) = \int E_\beta(\xi')\psi_{\boldsymbol{q}}^{*}(\xi')\,\mathrm{d}\xi'$$

so that

$$\begin{aligned} j_\alpha(\xi) = \hat{\sigma}_{\alpha\beta}E_\beta &= \int A_\beta(\boldsymbol{q})\hat{\sigma}_{\alpha\beta}\psi_{\boldsymbol{q}}(\xi)\,\mathrm{d}\boldsymbol{q} \\ &= \int \mathrm{d}\boldsymbol{q}\int \mathrm{d}\xi'\, E_\beta(\xi')\psi_{\boldsymbol{q}}^{*}(\xi')\hat{\sigma}_{\alpha\beta}\psi_{\boldsymbol{q}}(\xi). \end{aligned}$$

Using the notation

$$\sigma_{\alpha\beta}(\xi, \xi') = \int \mathrm{d}\boldsymbol{q}\, \psi_{\boldsymbol{q}}^{*}(\xi')\hat{\sigma}_{\alpha\beta}\psi_{\boldsymbol{q}}(\xi)$$

we obtain that

$$j_{\alpha}(\xi) = \int \sigma_{\alpha\beta}(\xi, \xi')E_{\beta}(\xi')\,\mathrm{d}\xi'.$$

This completes the proof.

Problem 5

The electromagnetic properties of a homogeneous isotropic medium without spatial dispersion may be described by the 'traditional' electric permittivity $\varepsilon(\omega)$ and the magnetic permeability $\mu(\omega)$. Express $\varepsilon(\omega)$ and $\mu(\omega)$ in terms of the limiting values of $\varepsilon_{\parallel}(k, \omega)$ and $\varepsilon_{\perp}(k, \omega)$ (see equation (2.11)) at $k \to 0$.

In terms of ε and μ the electric current in a medium is

$$\boldsymbol{j} = \partial \boldsymbol{P}/\partial t + c \operatorname{curl} \boldsymbol{M}$$

where $\boldsymbol{P}$ and $\boldsymbol{M}$ are the dielectric polarization and magnetization, which are equal to

$$\boldsymbol{P} = (\varepsilon - 1)\boldsymbol{E}/4\pi \qquad \boldsymbol{M} = (\mu - 1)\boldsymbol{B}/4\mu.$$

Using the Fourier representation and taking into account that $\boldsymbol{B} = c(\boldsymbol{k} \times \boldsymbol{E})/\omega$ (which follows from (2.3)) we find

$$\boldsymbol{j} = -\mathrm{i}\omega\boldsymbol{P} + \mathrm{i}c(\boldsymbol{k} \times \boldsymbol{M}) = \frac{-\mathrm{i}\omega(\varepsilon - 1)}{4\pi}\boldsymbol{E} + \frac{\mathrm{i}(\mu - 1)c^2}{4\pi\mu\omega}[\boldsymbol{k}(\boldsymbol{k}\cdot\boldsymbol{E}) - k^2\boldsymbol{E}].$$

Thus, the conductivity tensor

$$\sigma_{\alpha\beta} = \frac{-\mathrm{i}\omega(\varepsilon - 1)}{4\pi}\delta_{\alpha\beta} - \frac{\mathrm{i}(\mu - 1)k^2c^2}{4\pi\mu\omega}\left(\delta_{\alpha\beta} - \frac{k_\alpha k_\beta}{k^2}\right)$$

and, according to relation (2.8),

$$\varepsilon_{\alpha\beta} = \varepsilon\delta_{\alpha\beta} + \frac{(\mu - 1)k^2c^2}{\mu\omega^2}\left(\delta_{\alpha\beta} - \frac{k_\alpha k_\beta}{k^2}\right).$$

It is now seen from this expression and (2.11) that in the limit $k \to 0$, when spatial dispersion is absent,

$$\lim_{k\to 0} \varepsilon_{\parallel}(k, \omega) = \lim_{k\to 0} \varepsilon_{\perp}(k, \omega)$$

so

$$\varepsilon(\omega) = \lim_{k\to 0} \varepsilon_\perp(k, \omega) \qquad \mu^{-1}(\omega) = 1 - \frac{\omega^2}{c^2} \lim_{k\to 0} \frac{(\varepsilon_\perp - \varepsilon_\parallel)}{k^2}.$$

Problem 6

Calculate the tensor of the dielectric permeability for a 'cold' plasma formed by ions with infinitely large mass and electrons at rest with the density n (each electron has charge e and mass m), immersed in an external magnetic field with induction $\boldsymbol{B}_0$.

Since the ions are infinitely heavy, only electrons contribute to the current (the only role of ions is to compensate the volume charge of the electrons). Thus, in the linear approximation $\boldsymbol{j} = ne\boldsymbol{v}$, where $\boldsymbol{v}$ is the velocity of electrons which has to be determined from the linearized equation of motion:

$$m\frac{\partial \boldsymbol{v}}{\partial t} = e\boldsymbol{E} + \frac{e}{c}(\boldsymbol{v} \times \boldsymbol{B}_0).$$

Introducing the unit vector along the magnetic field $\boldsymbol{h} = \boldsymbol{B}_0/B_0$ and the electron gyrofrequency $\omega_B = eB_0/mc$, we obtain that

$$\boldsymbol{v} = \frac{\mathrm{i}e}{m\omega}\boldsymbol{E} + \frac{\mathrm{i}\omega_B}{\omega}(\boldsymbol{v} \times \boldsymbol{h}).$$

To find $\boldsymbol{v}$ it is convenient to obtain the scalar and cross products of this equation with $\boldsymbol{h}$:

$$(\boldsymbol{h}\cdot\boldsymbol{v}) = \frac{\mathrm{i}e}{m\omega}(\boldsymbol{h}\cdot\boldsymbol{E})$$

$$\boldsymbol{h} \times \boldsymbol{v} = \frac{\mathrm{i}e}{m\omega}(\boldsymbol{h} \times \boldsymbol{E}) + \frac{\mathrm{i}\omega_B}{\omega}[\boldsymbol{v} - \boldsymbol{h}(\boldsymbol{h}\cdot\boldsymbol{v})].$$

It helps to eliminate $[\boldsymbol{h} \times \boldsymbol{v}]$,

$$\boldsymbol{v} = \left(\frac{\mathrm{i}e}{m\omega}\boldsymbol{E} - \frac{e\omega_B}{m\omega^2}(\boldsymbol{h} \times \boldsymbol{E}) - \frac{\mathrm{i}e\omega_B^2}{m\omega^3}(\boldsymbol{h}\cdot\boldsymbol{E})\boldsymbol{h}\right)\left(1 - \frac{\omega_B^2}{\omega^2}\right)^{-1}.$$

Using the tensor notation we have for the conductivity tensor

$$\sigma_{\alpha\beta} = \left(1 - \frac{\omega_B^2}{\omega^2}\right)^{-1}\left(\frac{ine^2}{m\omega}\delta_{\alpha\beta} - \frac{ne^2\omega_B}{m\omega^2}e_{\alpha\beta\gamma}h_\gamma - \frac{ine^2\omega_B^2}{m\omega^3}h_\alpha h_\beta\right)$$

so, it follows from relation (2.8) that

$$\varepsilon_{\alpha\beta} = \left(1 - \frac{\omega_{pe}^2}{\omega^2 - \omega_B^2}\right)\delta_{\alpha\beta} - \frac{i\omega_{pe}^2\omega_B}{\omega(\omega^2 - \omega_B^2)} e_{\alpha\beta\gamma}h_\gamma + \frac{\omega_{pe}^2\omega_B^2}{\omega^2(\omega^2 - \omega_B^2)} h_\alpha h_\beta \tag{2.15}$$

where $\omega_{pe} = (4\pi ne^2/m)^{1/2}$ is the so-called electron plasma frequency.

Problem 7

Find the dispersion law for the electromagnetic waves in a 'cold' plasma without an external magnetic field.

According to (2.15) in the absence of the external magnetic field the dielectric permeability tensor for such a plasma has the simple form

$$\varepsilon_{\alpha\beta} = \varepsilon(\omega)\delta_{\alpha\beta} \qquad \varepsilon(\omega) = 1 - \omega_{pe}^2/\omega^2. \tag{2.16}$$

Therefore, as follows from the general rule (2.12), the dispersion equation for transverse waves is

$$k^2 = \frac{\omega_t^2}{c^2}\varepsilon(\omega_t) = \frac{\omega_t^2}{c^2}\left(1 - \frac{\omega_{pe}^2}{\omega_t^2}\right) \tag{2.17a}$$

that is

$$\omega_t(k) = (\omega_{pe}^2 + k^2c^2)^{1/2} \tag{2.17b}$$

For the longitudinal (electrostatic) oscillations $\varepsilon(\omega_l) = 1 - \omega_{pe}^2/\omega_l^2 = 0$, so $\omega_l = \omega_{pe}$. This is the so-called electron Langmuir oscillations in a plasma[†].

Problem 8

Each molecule of a rarefied gas is formed by two oppositely directed dipoles $\pm \boldsymbol{d}$, separated by the distance $\boldsymbol{a}$, which is proportional to the electric force $\mathscr{F}_\pm = \pm(\boldsymbol{d}\cdot\nabla)\boldsymbol{E}$ acting on dipoles: $\boldsymbol{a} = \gamma\mathscr{F}_+$. Determine the tensor of the dielectric permeability for such a gas, considering the orientation of molecules as chaotic. The density of gas is equal to n molecules per unit volume.

Let us start with some general procedure concerning the averaged (macroscopic) electric charges and currents in a medium. From a microscopic point

[†] More physical information about waves in a plasma may be found, for instance, in Stix T H 1962 *The Theory of Plasma Waves* (New York: McGraw-Hill).

of view the charge and current densities are as follows

$$\rho(\boldsymbol{r}) = \sum_{i,a} q_{ia}\delta(\boldsymbol{r} - \boldsymbol{r}_a - \boldsymbol{\xi}_{ia}) \qquad \boldsymbol{j}(\boldsymbol{r}) = \sum_{i,a} q_{ia}\dot{\boldsymbol{\xi}}_{ia}\delta(\boldsymbol{r} - \boldsymbol{r}_a - \boldsymbol{\xi}_{ia}).$$

Here a medium is represented as the number of atoms, where $\boldsymbol{r}_a$ is the position of the nucleus of the ath atom, and $\boldsymbol{\xi}_{ia}$ is the coordinate of the ith electron in this atom with the origin at the nucleus (so that $i = 0$ corresponds to the nucleus, i.e. $\boldsymbol{\xi}_{0a} = 0$, $q_{0a} = ze$). The macroscopic electric charge, $\langle \rho \rangle$, and current, $\langle \boldsymbol{j} \rangle$, are the result of the averaging of $\rho(\boldsymbol{r})$ and $\boldsymbol{j}(\boldsymbol{r})$, written above, over a 'physically infinitesimal' volume, which contains a large number of atoms, but which is small compared with the variation scale length λ of the macroscopic electromagnetic field. Since the size of an atom (which is of the order of ξ_i) is much less than λ from the very meaning of the macroscopic description, the typical values of $|\boldsymbol{r} - \boldsymbol{r}_a|$ will be much larger than ξ_i when this averaging is made. Therefore, the following expansion may be applied in calculating $\langle \rho \rangle$ and $\langle \boldsymbol{j} \rangle$:

$$\langle \rho \rangle \approx \left\langle \sum_{i,a} q_{ia}\delta(\boldsymbol{r} - \boldsymbol{r}_a) \right\rangle - \left\langle \sum_{i,a} q_{ia}\xi_{ia\alpha} \frac{\partial}{\partial x_\alpha} \delta(\boldsymbol{r} - \boldsymbol{r}_a) \right\rangle$$
$$+ \frac{1}{2}\left\langle \sum_{i,a} q_{ia}\xi_{ia\alpha}\xi_{ia\beta} \frac{\partial^2}{\partial x_\alpha\,\partial x_\beta} \delta(\boldsymbol{r} - \boldsymbol{r}_a) \right\rangle + \cdots$$
$$\langle j_\alpha \rangle \approx \left\langle \sum_{i,a} q_{ia}\dot{\xi}_{ia\alpha}\delta(\boldsymbol{r} - \boldsymbol{r}_a) \right\rangle - \left\langle \sum_{i,a} q_{ia}\dot{\xi}_{ia\alpha}\xi_{ia\beta} \frac{\partial}{\partial x_\beta} \delta(\boldsymbol{r} - \boldsymbol{r}_a) \right\rangle + \cdots.$$

The first term in the expansion for $\langle \rho \rangle$ is equal to zero because the atoms are neutral. The contribution of the second term may be written in the form

$$\langle \rho \rangle = -\operatorname{div} \boldsymbol{P} \qquad \boldsymbol{P} = \left\langle \sum_{i,a} q_{ia}\boldsymbol{\xi}_{ia}\delta(\boldsymbol{r} - \boldsymbol{r}_a) \right\rangle$$

where $\boldsymbol{P}$ is the dielectric polarization of the medium. However, it is easy to see that $\boldsymbol{P} = 0$ for the particular model considered in this problem. Hence we need to take into account the next term in the series. Then

$$\langle \rho \rangle = \frac{1}{2}\frac{\partial^2 \mathscr{D}_{\alpha\beta}}{\partial x_\alpha\,\partial x_\beta} \qquad \mathscr{D}_{\alpha\beta} = \left\langle \sum_{i,a} q_{ia}\xi_{ia\alpha}\xi_{ia\beta}\delta(\boldsymbol{r} - \boldsymbol{r}_a) \right\rangle$$

so in this case the averaged charge density is determined by the quadrupolar electric moment created in a medium.

It may now be verified straightforwardly that the expression for the macroscopic electric current $\langle \boldsymbol{j} \rangle$ can be transformed to the form

$$\langle \boldsymbol{j} \rangle = \frac{\partial \boldsymbol{P}}{\partial t} + c \operatorname{curl} \boldsymbol{M} - \frac{1}{2}\frac{\partial}{\partial t}\frac{\partial \mathscr{D}_{\alpha\beta}}{\partial x_\beta}$$

where

$$M = \left\langle \frac{1}{2c} \sum_{i,a} q_{ia}(\xi_{ia} \times \dot{\xi}_{ia})\delta(r - r_a) \right\rangle$$

is the magnetization of a medium. Thus knowing M and $\mathscr{D}_{\alpha\beta}$ we can find the conductivity tensor of the medium and, hence, its dielectric permeability. Let us calculate M and $\mathscr{D}_{\alpha\beta}$ for our model. It is convenient here to represent the dipole d as a limit of two opposite charges $\pm q$, separated by a distance ρ from each other, when $\rho \to 0$, $q \to \infty$, but $q\rho \to d$. The magnetic moment of a system of point charges

$$m = \frac{1}{2c} \sum_i q_i(r_i \times \dot{r}_i).$$

Therefore, it is easy to show that for two oppositely directed dipoles $\pm d$, separated by a distance a from each other, the magnetic moment is

$$m = \frac{1}{2c}(d \times \dot{a}).$$

Since for a molecule with a given value of d the distance $a = \gamma(d \cdot \nabla)E = i\gamma(d \cdot k)E$ (we assume that $E \propto e^{i(k \cdot r - \omega t)}$), it now follows that

$$m_\alpha = \frac{\gamma\omega}{2c} e_{\alpha\beta\mu} d_\beta d_\nu k_\nu E_\mu.$$

After the averaging of this expression over all possible directions of d (so that $\langle d_\beta d_\nu \rangle = \frac{1}{3} d^2 \delta_{\beta\nu}$) we obtain for M the following result:

$$M = \frac{\gamma n \omega d^2}{6c}(k \times E)$$

(n is the number of molecules per unit volume). A similar calculation can be done for $\mathscr{D}_{\alpha\beta}$. Starting from the quadrupolar moment for a system of point charges

$$D_{\alpha\beta} = \sum_i q_i x_{i\alpha} x_{i\beta}$$

we obtain step by step that

$$D_{\alpha\beta} = d_\alpha a_\beta + d_\beta a_\alpha = i\gamma(d_\alpha d_\mu k_\mu E_\beta + d_\beta d_\mu k_\mu E_\alpha)$$

and

$$\mathscr{D}_{\alpha\beta} = \frac{i\gamma n d^2}{3}(k_\alpha E_\beta + k_\beta E_\alpha).$$

Thus the macroscopic electric current

$$\langle j \rangle = c \operatorname{curl} \boldsymbol{M} - \frac{1}{2}\frac{\partial}{\partial t}\frac{\partial \mathscr{D}_{\alpha\beta}}{\partial x_\beta} = \mathrm{i}c(\boldsymbol{k} \times \boldsymbol{M}) - \frac{\omega}{2} k_\beta \mathscr{D}_{\alpha\beta} = -\frac{\mathrm{i}\omega\gamma n d^2 k^2}{3} E_\alpha.$$

Therefore, the conductivity tensor

$$\sigma_{\alpha\beta} = \frac{-\mathrm{i}\omega\gamma n d^2 k^2}{3} \delta_{\alpha\beta}$$

and

$$\varepsilon_{\alpha\beta} = \delta_{\alpha\beta} + \frac{4\pi \mathrm{i}}{\omega}\sigma_{\alpha\beta} = \left(1 + \frac{4\pi}{3}\gamma n d^2 k^2\right)\delta_{\alpha\beta}.$$

Thus, the permittivity of such a medium differs from unity only because of the spatial dispersion ($\boldsymbol{k} \neq 0$).

Problem 9

Determine the field of a point charge q in a uniaxial crystal.

A uniaxial crystal is an anisotropic dielectric whose permittivity, $\varepsilon_\parallel$, along the distinguished direction (called the optical axis) differs from its permittivity, $\varepsilon_\perp$, in the plane perpendicular to the optical axis (where it is isotropic). If n is the unit vector along the optical axis, the dielectric tensor for such a crystal may be written as follows:

$$\varepsilon_{\alpha\beta} = \varepsilon_\perp(\delta_{\alpha\beta} - n_\alpha n_\beta) + \varepsilon_\parallel n_\alpha n_\beta. \qquad (2.18)$$

Since $\boldsymbol{E} = -\nabla\varphi$, $\mathscr{D}_\alpha = \varepsilon_{\alpha\beta}E_\beta = -\varepsilon_{\alpha\beta}\,\partial\varphi/\partial x_\beta$, so

$$\operatorname{div}\mathscr{D} = -\frac{\partial}{\partial x_\alpha}\left(\varepsilon_{\alpha\beta}\frac{\partial\varphi}{\partial x_\beta}\right) = 4\pi q\delta(\boldsymbol{r}) \qquad (2.19)$$

(the charge being at the origin). Taking the coordinate z-axis along the optical axis we get from (2.18) and (2.19) that

$$\varepsilon_\parallel \frac{\partial^2\varphi}{\partial z^2} + \varepsilon_\perp\left(\frac{\partial^2\varphi}{\partial x^2} + \frac{\partial^2\varphi}{\partial y^2}\right) = -4\pi q\delta(x)\delta(y)\delta(z).$$

By the introduction of the new variable $z_* = z(\varepsilon_\perp/\varepsilon_\parallel)^{1/2}$ this becomes

$$\frac{\partial^2\varphi}{\partial x^2} + \frac{\partial^2\varphi}{\partial y^2} + \frac{\partial^2\varphi}{\partial z_*^2} = -\frac{4\pi q}{\varepsilon_\perp}\delta(x)\delta(y)\delta(z) = -\frac{4\pi q}{(\varepsilon_\perp\varepsilon_\parallel)^{1/2}}\delta(x)\delta(y)\delta(z_*).$$

This equation is the same as the usual Poisson equation for a point charge, so its solution is

$$\varphi(\boldsymbol{r}) = \frac{q}{r_*(\varepsilon_\perp \varepsilon_\parallel)^{1/2}} = \frac{q}{(\varepsilon_\parallel \varepsilon_\perp)^{1/2}[x^2 + y^2 + (\varepsilon_\perp/\varepsilon_\parallel)z^2]^{1/2}}.$$

Problem 10

Find the electrostatic potential of a point charge q in a medium with the dielectric tensor

$$\varepsilon_{\alpha\beta} = \varepsilon(k)\delta_{\alpha\beta} \qquad \varepsilon(k) = 1 + (ka)^{-2}.$$

By the Fourier transformation we obtain from (2.19) that

$$-k^2\varepsilon(k)\varphi(\boldsymbol{k}) = 4\pi\rho(\boldsymbol{k}) = \frac{4\pi q}{(2\pi)^{3/2}}\int \delta(\boldsymbol{r})\,\mathrm{e}^{-i\boldsymbol{k}\cdot\boldsymbol{r}}\,\mathrm{d}\boldsymbol{r} = \frac{4\pi q}{(2\pi)^{3/2}}.$$

Thus $\varphi(\boldsymbol{k}) = 2q/(2\pi)^{1/2}(k^2 + a^{-2})$, and therefore

$$\varphi(\boldsymbol{r}) = \frac{1}{(2\pi)^{3/2}}\int \varphi(\boldsymbol{k})\,\mathrm{e}^{i\boldsymbol{k}\cdot\boldsymbol{r}}\,\mathrm{d}\boldsymbol{k} = \frac{q}{\pi}\int_0^\pi \sin\theta\,\mathrm{d}\theta \int_0^\infty k^2\,\mathrm{d}k\,\frac{\mathrm{e}^{ikr\cos\theta}}{k^2 + a^{-2}} \tag{2.20}$$

(we use spherical coordinates in this integrand with the polar axis along $\boldsymbol{r}$). After a simple integration over the angle θ in (2.20)

$$\varphi(\boldsymbol{r}) = \frac{2q}{\pi r}\int_0^\infty \frac{\sin kr}{k^2 + a^{-2}}\,k\,\mathrm{d}k = \frac{2q}{\pi r}\int_0^\infty \frac{\sin(\kappa r/a)}{1+\kappa^2}\,\kappa\,\mathrm{d}\kappa. \tag{2.21}$$

To calculate this integral we use the odd and the even properties of the functions $\cos(\kappa r/a)$ and $\sin(\kappa r/a)$, so that

$$\int_0^\infty \frac{\sin\kappa z}{1+\kappa^2}\,\kappa\,\mathrm{d}\kappa = \frac{1}{2\mathrm{i}}\int_{-\infty}^{+\infty} \frac{\mathrm{e}^{\mathrm{i}\kappa z}}{1+\kappa^2}\,\kappa\,\mathrm{d}\kappa. \tag{2.22}$$

Now it is easy to find this with the help of the theory of residues in the complex plane κ (figure 2.1). Indeed, in the upper half-plane the integrand in (2.22) tends to zero exponentially. Thus, it is possible to supplement the integration over the real axis on (2.22) with the integral over the semicircle with a large enough radius in the upper half-plane, so

$$\int_{-\infty}^{+\infty} \frac{\mathrm{e}^{\mathrm{i}\kappa z}}{1+\kappa^2}\,\kappa\,\mathrm{d}\kappa = 2\pi\mathrm{i}\,\mathrm{Res}\left(\frac{\mathrm{e}^{\mathrm{i}\kappa z}}{1+\kappa^2}\,\kappa\right)_{\kappa=\mathrm{i}} = \mathrm{i}\pi\,\mathrm{e}^{-z}.$$

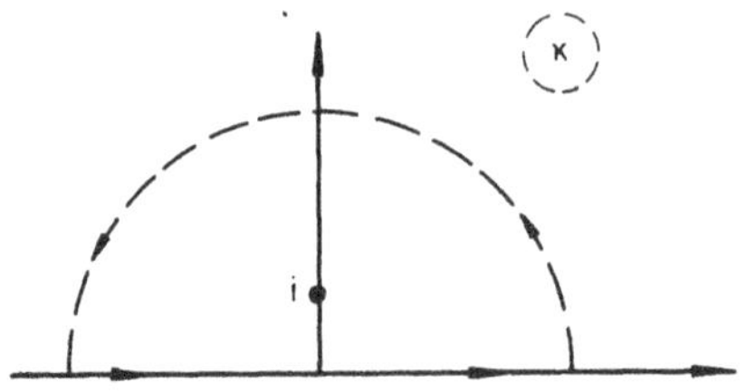

Figure 2.1. Contour of integration in the complex plane κ.

Finally, it follows from (2.21) that $\varphi(\boldsymbol{r}) = q\,\mathrm{e}^{-r/a}/r$. The exponential decrease of the potential at large distances ($r \gg a$) has a rather simple physical sense. The permittivity $\varepsilon(k)$ grows sharply for small values of k, $k \ll a^{-1}$, and this results in a strong screening of the Coulomb field (small k corresponds to large r in accordance to the uncertainty relation $kr \sim 1$). In reality such a screening takes place in a fully ionized gas (in a plasma)†.

Problem 11

A point charge q is immersed in a dielectric fluid (its permittivity is equal to ε) moving with the velocity $\boldsymbol{v}$. Determine the electrostatic potential, considering that $v < c/\varepsilon^{1/2}$.

It is convenient to determine at first the scalar and vector potentials of the electromagnetic field produced by this charge in the reference frame connected with the fluid, where the fluid is at rest and the charge is moving with a velocity equal to $-\boldsymbol{v}$. From the Maxwell equations

$$\operatorname{div}(\varepsilon \boldsymbol{E}) = 4\pi\rho_{\mathrm{ex}} \qquad \operatorname{div} \boldsymbol{B} = 0$$

$$\operatorname{curl} \boldsymbol{B} = \frac{4\pi}{c}\boldsymbol{j}_{\mathrm{ex}} + \frac{1}{c}\frac{\partial}{\partial t}(\varepsilon \boldsymbol{E}) \qquad \operatorname{curl} \boldsymbol{E} = -\frac{1}{c}\frac{\partial \boldsymbol{B}}{\partial t}$$

where $\rho_{\mathrm{ex}} = q\delta(\boldsymbol{r} + \boldsymbol{v}t)$ and $\boldsymbol{j}_{\mathrm{ex}} = -q\boldsymbol{v}\delta(\boldsymbol{r} + \boldsymbol{v}t)$, we find that

$$\boldsymbol{B} = \operatorname{curl} \boldsymbol{A} \qquad \boldsymbol{E} = -\nabla\varphi - \frac{1}{c}\frac{\partial \boldsymbol{A}}{\partial t}$$

and in the Lorentz gauge:

$$\operatorname{div} \boldsymbol{A} + \frac{\varepsilon}{c}\frac{\partial \varphi}{\partial t} = 0$$

† See, for instance, Landau L D and Lifshitz E M 1980 *Course of Theoretical Physics*, vol 5 (Oxford: Pergamon) part 1, §78.

the potentials A and φ satisfy the following wave equations:

$$\nabla^2 \boldsymbol{A} - \frac{\varepsilon}{c^2}\frac{\partial^2 \boldsymbol{A}}{\partial t^2} = -\frac{4\pi}{c}\boldsymbol{j}_{\text{ex}} \qquad \nabla^2 \varphi - \frac{\varepsilon}{c^2}\frac{\partial^2 \varphi}{\partial t^2} = -\frac{4\pi}{\varepsilon}\rho_{\text{ex}}$$

After the Fourier transformation it follows that

$$\begin{aligned} \boldsymbol{A}'(\boldsymbol{k}', \omega') &= \frac{-2q\boldsymbol{v}\delta(\omega' + \boldsymbol{k}'\cdot\boldsymbol{v})}{c\{(k')^2 - [(\omega')^2/c^2]\varepsilon\}} \\ \varphi'(\boldsymbol{k}', \omega') &= \frac{2q\delta(\omega' + \boldsymbol{k}\cdot\boldsymbol{v})}{\varepsilon\{(k')^2 - [(\omega')^2/c^2]\varepsilon\}} \end{aligned} \tag{2.23}$$

where the primes indicate that the corresponding quantities are related to the reference frame of the fluid. To go back to the laboratory frame (where the charge is at rest) we need to use the Lorentz transformation for the relativistic four vectors† $(\boldsymbol{A}, \mathrm{i}\varphi)$ and $(\boldsymbol{k}, \mathrm{i}\omega/c)$:

$$A_i = \alpha_{il}A'_l \qquad k_i = \alpha_{il}k'_l.$$

If $\boldsymbol{v}$ is directed along the x-axis the matrix α_{ik} is

$$\alpha_{ik} = \begin{pmatrix} \gamma & 0 & 0 & -\mathrm{i}\gamma v/c \\ 0 & 1 & 0 & 0 \\ 0 & 0 & 1 & 0 \\ \mathrm{i}\gamma v/c & 0 & 0 & \gamma \end{pmatrix}.$$

Then

$$\varphi(\boldsymbol{k}, \omega) = \gamma\left(\varphi' + \frac{v}{c}A'_x\right) \qquad A_x(\boldsymbol{k}, \omega) = \gamma\left(A'_x + \frac{v}{c}\varphi'\right)$$

$$A_y = A_z = 0$$

and

$$\omega' = \gamma(\omega - k_x v) \qquad k'_x = \gamma(k_x - v\omega/c^2) \qquad k'_y = k_y \qquad k'_z = k_z.$$

As a result it follows from (2.23) that

$$\varphi(\boldsymbol{k}, \omega) = \frac{2q\delta(\omega)}{\varepsilon(k_x^2(\gamma^2/\gamma_\varepsilon^2) + k_\perp^2)}\frac{\gamma^2}{\gamma_\varepsilon^2} \qquad A_x(\boldsymbol{k}, \omega) = -\frac{v(\varepsilon - 1)\gamma_\varepsilon^2}{c}\varphi(k, \omega)$$

where

$$\gamma^2 = (1 - v^2/c^2)^{-1} \qquad \gamma_\varepsilon^2 = (1 - \varepsilon v^2/c^2)^{-1}$$

(we put $\omega = 0$ in these expressions since φ and A_x are proportional to $\delta(\omega)$).

† See, Landau L D and Lifshitz E M 1971 *Course of Theoretical Physics*, vol 2 (Oxford: Pergamon) §6.

In the $(\boldsymbol{r}, t)$ representation we now have for $\varphi(\boldsymbol{r}, t)$:

$$\varphi(\boldsymbol{r}, t) = \frac{1}{(2\pi)^2}\int \varphi(\boldsymbol{k}, \omega)\,\mathrm{e}^{\mathrm{i}(\boldsymbol{k}\cdot\boldsymbol{r} - \omega t)}\,\mathrm{d}\boldsymbol{k}\,\mathrm{d}\omega$$

$$= (2\pi)^{-2}\int \frac{2q\gamma^2/\gamma_\varepsilon^2}{\varepsilon[(k_x^2\gamma^2/\gamma_\varepsilon^2) + k_\perp^2]}\,\mathrm{e}^{\mathrm{i}\boldsymbol{k}\cdot\boldsymbol{r}}\,\mathrm{d}\boldsymbol{k}. \qquad (2.24)$$

This integral may be calculated in the same way that has been used in problem 3 (see formula (1.10)). By the introduction of the new variables, $\boldsymbol{k}_* = (k_x\gamma/\gamma_\varepsilon, \boldsymbol{k}_\perp)$ and $\boldsymbol{r}_* = (x\gamma_\varepsilon/\gamma, \boldsymbol{r}_\perp)$, the integrand in (2.24) becomes similar to that in (1.10). Thus

$$\varphi(\boldsymbol{r}) = \frac{q\gamma/\gamma_\varepsilon}{\varepsilon[(x^2\gamma_\varepsilon^2/\gamma^2) + y^2 + z^2]^{1/2}}.$$

It is worth noting here that such a procedure is possible only under the condition that $\gamma_\varepsilon^2 > 0$, that is $v < c/\varepsilon^{1/2}$. If the velocity of a fluid exceeds this critical value (which is equal to the propagation velocity of the electromagnetic waves in this medium), the electromagnetic field differs from zero only inside the 'Cherenkov cone' with the angle $\cos\theta = c/v\varepsilon^{1/2}$†.

Comparing this expression for the scalar potential with the solution of problem 9, it is easy to see that from the electrostatic point of view such a moving dielectric fluid is analogous to a uniaxial crystal with the optical axis along $\boldsymbol{v}$ and the permittivities $\varepsilon_\parallel = \varepsilon$ and $\varepsilon_\perp = \varepsilon\gamma_\varepsilon^2/\gamma^2$. However, the important difference is that in this case the magnetic field is also produced:

$$A_x(\boldsymbol{r}) = -\frac{v(\varepsilon - 1)}{c}\gamma_\varepsilon^2\varphi(\boldsymbol{r})$$

so

$$\boldsymbol{B}(\boldsymbol{r}) = \operatorname{curl}\boldsymbol{A} = (\varepsilon - 1)\gamma_\varepsilon^2(\boldsymbol{E}\times\boldsymbol{v}/c).$$

Let us note that this problem may be solved without transition to the reference frame of the fluid by using Minkowski's formulae for moving dielectrics‡.

Problem 12

Determine the dispersion equations for electromagnetic waves in a uniaxial crystal.

† See, Ginzburg V L 1989 *Applications of Electrodynamics in Theoretical Physics and Astrophysics* (New York: Gordon and Breach) ch. 7.

‡ See, Landau L D and Lifshitz E M 1984 *Course of Theoretical Physics* vol 8 (Oxford: Pergamon) §76.

Let us use the coordinate system with the z-axis directed along the optical axis of a crystal and the wave vector $\boldsymbol{k}$ lying in the x–z plane: $\boldsymbol{k} = (k_x, 0, k_z)$. Then for the tensor $\varepsilon_{\alpha\beta}$ given by (2.18) the equations (2.9) result in

$$\left(-k_z^2 + \frac{\omega^2}{c^2}\varepsilon_\perp\right)E_x + k_x k_z E_z = 0$$

$$\left(-k^2 + \frac{\omega^2}{c^2}\varepsilon_\perp\right)E_y = 0 \tag{2.25}$$

$$\left(-k_x^2 + \frac{\omega^2}{c^2}\varepsilon_\parallel\right)E_z + k_z k_x E_x = 0.$$

The corresponding dispersion equation: det $\| L_{\alpha\beta} \| = 0$, takes the form

$$\left(-k^2 + \frac{\omega^2}{c^2}\varepsilon_\perp\right)\left[\left(-k_z^2 + \frac{\omega^2}{c^2}\varepsilon_\perp\right)\left(-k_x^2 + \frac{\omega^2}{c^2}\varepsilon_\parallel\right) - k_x^2 k_z^2\right] = 0$$

with two solutions. The first is: $-k^2 + (\omega^2/c^2)\varepsilon_\perp = 0$. Such a wave is called the *ordinary wave*, since its refraction index $n = \varepsilon_\perp^{1/2}$ has no dependence on the direction of $\boldsymbol{k}$. It is seen from (2.25) that this wave is linearly polarized; the electric field vector has only a y-component. The magnetic induction vector $\boldsymbol{B}$, is in the x–z plane and is directed perpendicularly to $\boldsymbol{k}$. The second solution may be written as

$$\frac{k_x^2}{\varepsilon_\parallel} + \frac{k_z^2}{\varepsilon_\perp} = \frac{\omega^2}{c^2}$$

and corresponds to the *extraordinary wave*. If θ is the angle between the wave vector $\boldsymbol{k}$ and the optical axis, we obtain for the refractive index

$$n = \left(\frac{\sin^2\theta}{\varepsilon_\parallel} + \frac{\cos^2\theta}{\varepsilon_\perp}\right)^{-1/2}$$

so for the extraordinary wave this is not isotropic (unlike the ordinary wave). This wave is also linearly polarized (which follows from (2.25)), but now the vector $\boldsymbol{B}$ is directed along the y-axis, while $\boldsymbol{E}$ is in x–z plane.

Problem 13

The angle between the wave vector $\boldsymbol{k}$ of a extraordinary electromagnetic wave and the optical axis of a uniaxial crystal (with given $\varepsilon_\parallel$ and $\varepsilon_\perp$) is equal to θ. Find the direction of the ray vector for this wave.

We choose, as in the previous problem, the coordinate system with the z-axis along the optical axis and the vector $\boldsymbol{k}$ in the x–z plane. So for the

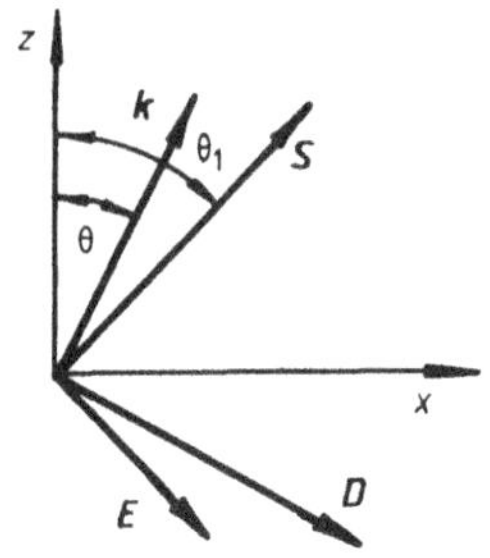

Figure 2.2. Geometry of the extraordinary electromagnetic wave.

extraordinary wave vector $\boldsymbol{B}$ is perpendicular to this plane, and the vectors $\boldsymbol{E}$ *and* $\mathscr{D}$ are in it. Since the electric induction vector $\mathscr{D}$ is transverse to $\boldsymbol{k}$ (see figure 2.2)

$$\mathscr{D}_x = \mathscr{D}\cos\theta \qquad \mathscr{D}_z = -\mathscr{D}\sin\theta.$$

Therefore, the components of the electric field are

$$E_x = \mathscr{D}_x/\varepsilon_\perp = \mathscr{D}\cos\theta/\varepsilon_\perp \qquad E_z = \mathscr{D}_z/\varepsilon_\parallel = -\mathscr{D}\sin\theta/\varepsilon_\parallel.$$

The direction of the ray vector is determined by the vector of the energy flux, that is by the Poynting vector $\boldsymbol{S} = (c/4\pi)(\boldsymbol{E}\times\boldsymbol{B})$. Thus the directions of the ray and wave vectors are not the same for the extraordinary wave. It is seen from the figure 2.2 that the ray vector is coplanar with the optical axis and $\boldsymbol{k}$, so that the angle θ_1 between $\boldsymbol{S}$ and the optical axis is

$$\tan\theta_1 = -E_z/E_x = (\varepsilon_\perp/\varepsilon_\parallel)\tan\theta$$

(θ_1 is equal to θ only for $\theta = 0, \pi/2$).

Problem 14

Determine the reflection coefficient for a linearly polarized electromagnetic wave incident at an angle θ_0 from a vacuum on the plane boundary of a dielectric medium with the given permittivity ε and the magnetic permeability μ (figure 2.3).

Let the boundary be the $z = 0$ plane, and the plane of incidence be the x–z plane. The geometry of reflection and refraction is completely determined by the homogeneity of the system in the x–y plane and the dispersion relations for incident (0), reflected (1) and refracted (2) waves. Because of this homogeneity the components k_x and k_y of the wave vector must be the same for all three waves. Thus, $k_{1y} = k_{2y} = k_{0y} = 0$, and all the waves propagate in the plane of incidence. Since

$$k_0 = k_1 = \omega/c \qquad k_2 = (\omega/c)(\varepsilon\mu)^{1/2}$$

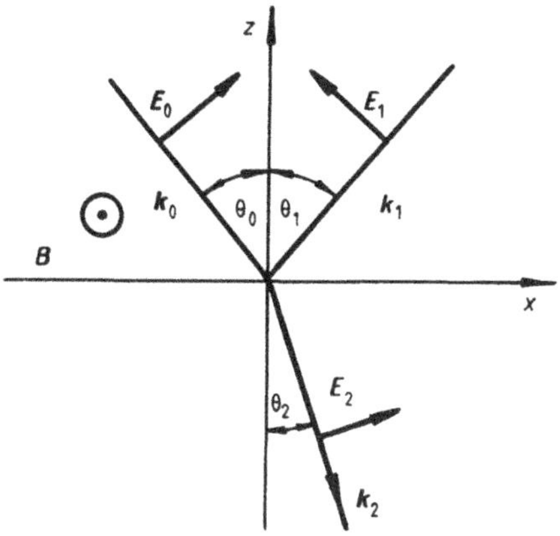

Figure 2.3. Reflection and refraction of the electromagnetic wave.

(the frequency ω is also the same for all the waves) it follows from the equations

$$k_{1x} = k_{2x} = k_{0x}$$

that $\theta_1 = \theta_0$ and

$$\sin\theta_2 = (\sin\theta_0)/(\varepsilon\mu)^{1/2}.$$

The reflection (and refraction) coefficient depends on the polarization of the incident wave. We consider here the particular case (see figure 2.3) when the vector $\boldsymbol{B}$ is perpendicular to the plane of incidence (the directions of $\boldsymbol{E}$ for all the waves are shown). The amplitudes of the reflected and refracted waves may be determined from the boundary conditions at the surface of separation ($z = 0$), namely from the continuity of the tangential components of the vectors $\boldsymbol{E}$ and $\boldsymbol{H} = \boldsymbol{B}/\mu$:

$$B_0 + B_1 = B_2/\mu$$

$$E_0\cos\theta_0 - E_1\cos\theta_0 = E_2\cos\theta_2.$$

Taking into account that $B_0 = E_0, B_1 = E_1$ and $B_2 = E_2(\varepsilon\mu)^{1/2}$ we find that

$$\begin{aligned} E_1 &= E_0\frac{(\cos\theta_0)(\varepsilon/\mu)^{1/2} - [1 - (\sin^2\theta_0)/\varepsilon\mu]^{1/2}}{(\varepsilon/\mu)^{1/2}\cos\theta_0 + [1 - (\sin^2\theta_0)/\varepsilon\mu]^{1/2}} \\ E_2 &= E_0\frac{2\cos\theta_0}{(\varepsilon/\mu)^{1/2}\cos\theta_0 + [1 - (\sin^2\theta_0)/\varepsilon\mu]^{1/2}}. \end{aligned} \tag{2.26}$$

The reflection coefficient

$$R = |E_1/E_0|^2.$$

Problem 15

Find the condition under which the surface electromagnetic wave can propagate along the plane boundary between the vacuum and a dielectric of given permittivity $\varepsilon(\omega)$. Obtain the corresponding dispersion relation for the particular case of the electron plasma with $\varepsilon(\omega) = 1 - \omega_{\mathrm{pe}}^2/\omega^2$ (see problem 7).

We place a dielectric in the domain $z < 0$ and the wave propagates along the x-axis. Thus, the dependence of the field on space and time is proportional to $f(z)\,\mathrm{e}^{\mathrm{i}(kx-\omega t)}$, where ω is the frequency and k is the wave vector. From the Maxwell equations

$$\operatorname{curl} \boldsymbol{B} = \frac{1}{c}\frac{\partial \mathscr{D}}{\partial t} = -\frac{\mathrm{i}\omega}{c}\varepsilon \boldsymbol{E}$$

$$\operatorname{curl} \boldsymbol{E} = -\frac{1}{c}\frac{\partial \boldsymbol{B}}{\partial t} = \frac{\mathrm{i}\omega}{c}\boldsymbol{B}$$

we have for the electric field $\boldsymbol{E}$:

$$\nabla^2 \boldsymbol{E} + \frac{\omega^2}{c^2}\varepsilon \boldsymbol{E} = 0.$$

We need to solve this equation separately in the vacuum ($z > 0$) and in the dielectric ($z < 0$). Since we are interested in the solution that tends to zero while moving off the plane $z = 0$ (the surface wave!) we find

$$\boldsymbol{E} = \begin{cases} \boldsymbol{E}_1 \exp(-\kappa_1 z)\exp[\mathrm{i}(kx - \omega t)] & z > 0 \quad \kappa_1 = (k^2 - \omega^2/c^2)^{1/2} > 0. \\ \boldsymbol{E}_2 \exp(\kappa_2 z)\exp[\mathrm{i}(kx - \omega t)] & z < 0 \quad \kappa_2 = (k^2 - \omega^2\varepsilon/c^2)^{1/2} > 0. \end{cases} \tag{2.27}$$

To determine the polarization of a surface electromagnetic wave and its dispersion relation let us write the Maxwell equations in their component form:

$$-\frac{\partial E_y}{\partial z} = \frac{\mathrm{i}\omega}{c}B_x \qquad -\frac{\partial B_y}{\partial z} = -\frac{\mathrm{i}\omega}{c}\varepsilon E_x$$

$$\frac{\partial E_x}{\partial z} - \mathrm{i}kE_z = \frac{\mathrm{i}\omega}{c}B_y \qquad \frac{\partial B_x}{\partial z} - \mathrm{i}kB_z = -\frac{\mathrm{i}\omega}{c}\varepsilon E_y \tag{2.28}$$

$$\mathrm{i}kE_y = \frac{\mathrm{i}\omega}{c}B_z \qquad \mathrm{i}kB_y = -\frac{\mathrm{i}\omega}{c}\varepsilon E_z.$$

Since E_y and B_x are continuous across the boundary it follows from the first equation in the left-hand column of (2.28) that $E_y = B_x = 0$. Indeed if this were not the case the left-hand side of this equation ($\partial E_y/\partial z$) would be discontinuous (see (2.27)) while the right-hand side would not be. Then the third equation in the left-hand column of (2.28) tells us that $B_z = 0$. Thus, only the B_y, E_x and E_z components are present in this wave. Let us now write the first equation in the right-hand column of (2.28) in a vacuum and in a dielectric:

$$\kappa_1 B_y = -\frac{i\omega}{c} E_x \qquad \kappa_2 B_y = \frac{i\omega}{c}\varepsilon(\omega) E_x.$$

Since the components B_y and E_x are continuous at the boundary, the solvability condition is $\kappa_2 = -\varepsilon(\omega)\kappa_1$. Thus, such a surface wave may exist only in the frequency range where the permittivity of a dielectric $\varepsilon(\omega)$ is negative. By substitution of the expressions (2.27) for κ_1 and κ_2 we obtain the following dispersion equation:

$$[k^2 - \omega^2\varepsilon(\omega)/c^2]^{1/2} = -\varepsilon(\omega)(k^2 - \omega^2/c^2)^{1/2}. \qquad (2.29)$$

For the particular case of the electron plasma with $\varepsilon(\omega) = 1 - \omega_{pe}^2/\omega^2$ it follows simply from (2.29) that

$$\omega(k) = \omega_{pe}\{1 + 2(kc/\omega_{pe})^2 - [1 + 4(kc/\omega_{pe})^4]^{1/2}\}^{1/2}/\sqrt{2}.$$

A qualitative plot of this curve is shown in figure 2.4. For the limiting case of short wavelength, when $kc/\omega_{pe} \gg 1$, such a surface electromagnetic

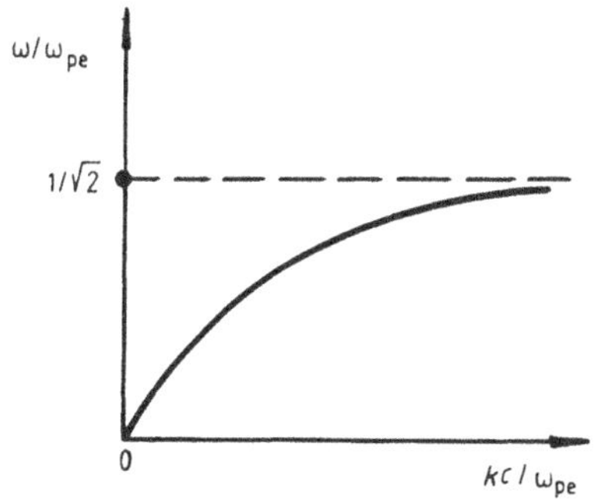

Figure 2.4. Dispersion law for the surface electromagnetic wave.

wave becomes almost potential (electrostatic) so that $\boldsymbol{B}$ is much less than $\boldsymbol{E}$. The dispersion equation in this limit takes the simple form: $\varepsilon(\omega) = -1$. This follows at once from the general dispersion relation (2.29) by putting $c \to \infty$.

Problem 16

Determine the mean values of the electromagnetic part of the internal energy and momentum per unit volume of a transparent medium.

This part of the energy and momentum densities may be designated as the difference of these quantities in a medium with and without an electromagnetic wave (under other fixed macroscopic parameters of a medium, for instance, the density and the temperature). This difference may be calculated in the following way. Suppose that a small amount of dissipation (and corresponding damping of the wave) is present. Then the amplitude of the wave is a decreasing function that tends to zero as $t \to \infty$.

At this process the densities of the energy (W) and of the momentum ($\boldsymbol{P}$) of the electromagnetic wave decrease owing to their absorption in a medium:

$$\mathrm{d}W/\mathrm{d}t = -Q \qquad \mathrm{d}\boldsymbol{P}/\mathrm{d}t = -\boldsymbol{\Pi},$$

where Q and $\boldsymbol{\Pi}$ are the powers of the energy and the momentum absorptions per unit volume. By the integration of these relations over time from the initial moment and up to infinity, when the wave is absent, we obtain

$$W = \int_0^\infty Q\,\mathrm{d}t \qquad \boldsymbol{P} = \int_0^\infty \boldsymbol{\Pi}\,\mathrm{d}t. \tag{2.30}$$

The energy absorption $Q = \langle \boldsymbol{j}\cdot\boldsymbol{E}\rangle$ (the sign $\langle\ \rangle$ means the average over the period of the wave). At first we calculate this quantity for a monochromatic field of given frequency ω. Since Q is proportional to the second power of the field amplitude we need to write $\boldsymbol{j}$ and $\boldsymbol{E}$ in the real form:

$$\boldsymbol{E} = \tfrac{1}{2}(\boldsymbol{E}_0\,\mathrm{e}^{-\mathrm{i}\omega t} + \boldsymbol{E}_0^*\,\mathrm{e}^{\mathrm{i}\omega t}) \qquad \boldsymbol{j} = \tfrac{1}{2}(\boldsymbol{j}_0\,\mathrm{e}^{-\mathrm{i}\omega t} + \boldsymbol{j}_0^*\,\mathrm{e}^{\mathrm{i}\omega t})$$

so that

$$\begin{aligned} Q = \langle \boldsymbol{j}\cdot\boldsymbol{E}\rangle &= \tfrac{1}{4}(E_0 j_0^* + E_0^* j_0) = \tfrac{1}{4}(E_{0\alpha}\sigma_{\alpha\beta}^* E_{0\beta}^* + E_{0\beta}^*\sigma_{\beta\alpha}E_{0\alpha}) \\ &= \tfrac{1}{4}E_{0\alpha}E_{0\beta}^*(\sigma_{\beta\alpha} + \sigma_{\alpha\beta}^*) = -\frac{\mathrm{i}\omega}{16\pi}(\varepsilon_{\alpha\beta} - \varepsilon_{\beta\alpha}^*)E_{0\alpha}^*E_{0\beta} \end{aligned}$$

(the relation (2.8) for $\sigma_{\alpha\beta}$ and $\varepsilon_{\alpha\beta}$ has been used here).

Thus the power dissipated in a medium is determined by the antiHermitian part (see formula (1.4)) of the dielectric permeability tensor:

$$Q = \frac{-\mathrm{i}\omega}{8\pi}\varepsilon_{\alpha\beta}^{(\mathrm{A})}E_{0\alpha}^*E_{0\beta}. \tag{2.31}$$

This dissipation leads to the damping of the wave with some decrement γ, so the amplitude of the wave has the form $\boldsymbol{E} = \boldsymbol{E}_0\,\mathrm{e}^{-\gamma t}$. We assume here and in what follows that this damping is weak, so that $\gamma \ll \omega$ (or, what is the same thing, $\varepsilon_{\alpha\beta}^{(\mathrm{A})} \ll \varepsilon_{\alpha\beta}^{(\mathrm{H})}$). For a given value of γ the dissipated power Q is proportional to $\mathrm{e}^{-2\gamma t}$, thus it follows from (2.30) and (2.31) after integration that

$$W = \frac{-\mathrm{i}\omega}{16\pi\gamma}\varepsilon_{\alpha\beta}^{(\mathrm{A})}E_{0\alpha}^*E_{0\beta}. \tag{2.32}$$

It is clear from the very designation of W that this quantity must be independent of the damping rate, while the latter is present in (2.32). However, this may really be excluded by taking into account the relation between γ and $\varepsilon_{\alpha\beta}^{(A)}$, which follows from the dispersion equations (2.9) and (2.10). In the absence of damping the tensor $\varepsilon_{\alpha\beta}$ is a Hermitian one, $\gamma = 0$, that is the frequency ω is purely real, and the electric field amplitude $\boldsymbol{E}_0$ satisfies equation (2.9):

$$L_{\alpha\beta}E_\beta = (k_\alpha k_\beta - k^2\delta_{\alpha\beta} + \omega^2\varepsilon_{\alpha\beta}^{(H)}/c^2)E_{0\beta} = 0.$$

If a small but finite value of $\varepsilon_{\alpha\beta}^{(A)}$ is present, the small imaginary correction to the frequency of a wave appears: $\delta\omega = -\mathrm{i}\gamma$, so the operator $L_{\alpha\beta}$ takes the form:

$$L_{\alpha\beta}(\omega - \mathrm{i}\gamma) \approx L_{\alpha\beta}(\omega) - \mathrm{i}\gamma\,\partial L_{\alpha\beta}/\partial\omega = L_{\alpha\beta}^{(H)} + L_{\alpha\beta}^{(A)}$$

$$L_{\alpha\beta}^{(H)} = k_\alpha k_\beta - k^2\delta_{\alpha\beta} + \omega^2\varepsilon_{\alpha\beta}^{(H)}/c^2$$

$$L_{\alpha\beta}^{(A)} = \frac{\omega^2\varepsilon_{\alpha\beta}^{(A)}}{c^2} - \frac{\mathrm{i}\gamma}{c^2}\frac{\partial}{\partial\omega}(\omega^2\varepsilon_{\alpha\beta}^{(H)}).$$

The eigenvector E is also slightly different now: $E_\beta = E_{0\beta} + \delta E_\beta$, and is governed by the equation (2.9)

$$L_{\alpha\beta}E_\beta = (L_{\alpha\beta}^{(H)} + L_{\alpha\beta}^{(A)})(E_{0\beta} + \delta E_\beta) \approx L_{\alpha\beta}^{(H)}\delta E_\beta + L_{\alpha\beta}^{(A)}E_{0\beta} = 0. \quad (2.33)$$

Since we are now interested only in the decrement γ rather than in δE_β (the latter describes the small correction to the polarization of the wave) let us take the scalar product of equation (2.33) with the vector $E_{0\alpha}^*$:

$$L_{\alpha\beta}^{(A)}E_{0\alpha}^*E_{0\beta} = -L_{\alpha\beta}^{(H)}\delta E_\beta E_{0\alpha}^* = -L_{\beta\alpha}^{(H)*}E_{0\alpha}^*\delta E_\beta = 0$$

(we take into account here that $L_{\beta\alpha}^{(H)*} = L_{\alpha\beta}^{(H)}$, and also that $E_{0\alpha}$ is the eigenvector of this operator: $L_{\alpha\beta}^{(H)}E_{0\beta} = 0$).

As a result we have that

$$\omega^2\varepsilon_{\alpha\beta}^{(A)}E_{0\alpha}^*E_{0\beta} = \mathrm{i}\gamma\frac{\partial}{\partial\omega}(\omega^2\varepsilon_{\alpha\beta}^{(H)})E_{0\alpha}^*E_{0\beta}.$$

By substitution of this expression into (2.32) we finally obtain that

$$W = \frac{1}{16\pi\omega}\frac{\partial}{\partial\omega}(\omega^2\varepsilon_{\alpha\beta}^{(H)})E_{0\alpha}^*E_{0\beta}.$$

To find the momentum density of a wave $\boldsymbol{P}$ let us note that the absorbed momentum $\boldsymbol{\Pi}$ is equal to the mean force acting on the medium (per unit volume). Therefore,

$$\boldsymbol{\Pi} = \langle \rho\boldsymbol{E} + c^{-1}(\boldsymbol{j} \times \boldsymbol{B})\rangle.$$

Since $\boldsymbol{B} = c(\boldsymbol{k} \times \boldsymbol{E})/\omega$, and the charge density ρ is linked with the current $\boldsymbol{j}$ by the continuity equation: $\partial\rho/\partial t + \operatorname{div}\boldsymbol{j} = 0$, or in the Fourier transform representation, $\rho = (\boldsymbol{k}\cdot\boldsymbol{j})/\omega$, we have representation, $\rho = (\boldsymbol{k}\cdot\boldsymbol{j})/\omega$, we have

$$\Pi = \frac{\boldsymbol{k}}{\omega}\langle \boldsymbol{j}\cdot\boldsymbol{E}\rangle = \boldsymbol{k}Q/\omega.$$

Thus it follows at once that $\boldsymbol{P} = \boldsymbol{k}W/\omega$.

Problem 17

Check the general expressions (2.14) for the energy and momentum of an electromagnetic wave in a medium by the straightforward calculation of these quantities for the electron Langmuir oscillations (see problem 7).

We shall use the designation of the energy and momentum of a wave given in the previous problem. In the absence of oscillations the electric field is equal to zero, and the electrons (and ions) are at rest, so the energy and the momentum of the medium are equal to zero. If the Langmuir wave propagating along the z-axis is present we have for the electric field:

$$\boldsymbol{E} = \boldsymbol{e}_z E_0 \cos(kz - \omega_{\text{pe}}t).$$

The energy of the medium is now formed by the energy of the electric field and the kinetic energy of the electrons (the heavy ions are still at rest):

$$W = \frac{\langle E^2\rangle}{8\pi} + \frac{nm}{2}\langle v^2\rangle.$$

From the equation of motion (its z-component) for the electrons

$$m\,\partial v/\partial t = eE_z$$

their velocity is

$$v = -eE_0 \sin(kz - \omega t)/m\omega_{\text{pe}}.$$

Thus, $W = E_0^2/8\pi$. Let us calculate the momentum. The momentum density for the electromagnetic field† $(\boldsymbol{E}\times\boldsymbol{B})/4\pi c$ is equal to zero in this case ($\boldsymbol{B} = 0$, the electrostatic oscillations). Thus only electrons contribute to the momentum of a wave,

$$\boldsymbol{P} = \langle mv(n + \delta n)\rangle \boldsymbol{e}_z = m\langle v\delta n\rangle \boldsymbol{e}_z.$$

Here δn is the perturbation of the electron density in a wave: this may be determined from the linearized continuity equation

$$\partial\delta n/\partial t + n\,\partial v/\partial x = 0$$

† See, Landau L D and Lifshitz E M 1971 *Course of Theoretical Physics*, vol 2 (Oxford: Pergamon) §4.

so

$$\delta n = -\frac{kneE_0}{m\omega_{\text{pe}}^2}\sin(kz - \omega t).$$

As a result

$$\boldsymbol{P} = \boldsymbol{e}_z m\langle \delta n v\rangle = \boldsymbol{e}_z \frac{kE_0^2}{8\pi\omega_{\text{pe}}} = \boldsymbol{k}W/\omega_{\text{pe}}$$

(the relation $\omega_{\text{pe}}^2 = 4\pi ne^2/m$ has been used here). Thus the expression for the momentum is the same as in (2.14). The dielectric tensor for such a plasma is (see (2.16)): $\varepsilon_{\alpha\beta} = \delta_{\alpha\beta}(1 - \omega_{\text{pe}}^2/\omega^2)$. Thus, according to (2.14),

$$W = \frac{E_0^2}{16\pi\omega}\frac{\partial}{\partial\omega}[\omega^2\varepsilon(\omega)]_{\omega=\omega_{\text{pe}}} = \frac{E_0^2}{8\pi}$$

which also coincides with the result of straightforward calculation.

Problem 18

Show that in the reference frame moving relative to a plasma with the velocity $u > \omega_{\text{pe}}/k$, the energy of the Langmuir wave, considered in the previous problem, becomes negative.

Let us write the linearized Poisson equation, the continuity equation and the equation of motion for the electrons (taking into account that in the unperturbed state the latter move with the velocity u):

$$\partial E/\partial z = 4\pi e\delta n$$

$$\frac{\partial \delta n}{\partial t} + \frac{\partial}{\partial z}(n\delta v + u\delta n) = 0$$

$$\frac{\partial \delta v}{\partial t} + u\frac{\partial \delta v}{\partial z} = eE/m.$$

Searching for a solution in the form of a wave propagating along the z-axis, when $E = E_0\cos(kz - \omega t)$, we find for the velocity and density perturbations:

$$\delta n = -\frac{eE_0kn}{\omega_{\text{pe}}^2 m}\sin(kz - \omega t) \qquad \delta v = \frac{-eE_0}{m(\omega - ku)}\sin(kz - \omega t)$$

where the dispersion equation takes the form: $(\omega - ku)^2 = \omega_{\text{pe}}^2$, that is $\omega = ku \pm \omega_{\text{pe}}$. This describes the Doppler shift of the frequency, and two signs $(\pm\omega_{\text{pe}})$ correspond to waves that propagate to the right or to the left along the z-axis in the frame where plasma is at rest. Let us consider the wave that propagates backwards in this frame, for which $\omega - ku = -\omega_{pe}$, and calculate its energy. The energy of the medium with such a wave is formed by the energy of the electric field $W_e = \langle E^2 \rangle / 8\pi = E_0^2/16\pi$ and by the kinetic energy of the electrons

$$W_k = \langle \tfrac{1}{2}m(n + \delta n)(u + \delta v)^2 \rangle = \tfrac{1}{2}nmu^2 + \tfrac{1}{2}mn\langle (\delta v)^2 \rangle + mu\langle \delta n \delta v \rangle.$$

According to the designation given above the energy of a wave is $W = W_e + W_k - \frac{1}{2}mnu^2$ (we subtract the energy of this medium without a wave). It is easy to obtain after simple calculations that $W = E_0^2(1 - ku/\omega_{pe})/8\pi$. Thus, the energy of such a wave indeed becomes negative for $u > \omega_{pe}/k$.

Problem 19

Determine the energy of the surface electromagnetic wave considered in problem 15.

We may use the same procedure as in problem 16. Thus we consider that a small antiHermitian part is present in the dielectric tensor of the medium and calculate the corresponding damping of the wave. Since we deal here with the isotropic dielectric without spatial dispersion, when $\varepsilon_{\alpha\beta} = \varepsilon(\omega)\delta_{\alpha\beta}$, its antiHermitian part is determined by the imaginary part of the permittivity $\varepsilon(\omega)$: $\varepsilon_{\alpha\beta}^{(A)} = i\varepsilon''(\omega)\delta_{\alpha\beta}$. Therefore, according to (2.31), the dissipated power per unit volume is $Q = \omega\varepsilon''(\omega)|E|^2/8\pi$. For the surface wave the energy per unit area (rather than per unit volume) has physical meaning. Thus we need to integrate Q over z from zero to $-\infty$, that is, over the volume of a dielectric per unit boundary area. Since the amplitude of the electric field decreases exponentially inside a dielectric as $\exp(\kappa_2 z)$ then $Q \propto \exp(2\kappa_2 z)$ and so

$$\tilde{Q} = \int_{-\infty}^{0} Q \, dz = \omega\varepsilon''(\omega)|E_0|^2/16\pi\kappa_2$$

where $|E_0|$ is the amplitude of the electric field in a dielectric at the boundary ($z = 0$). This dissipation results in the damping of a wave with the decrement γ, so, in analogy with (2.30)

$$\tilde{W} = \int_0^{\infty} \tilde{Q} \, dt = \frac{\tilde{Q}_0}{2\gamma} = \frac{|E_0|^2\omega\varepsilon''}{(16\pi\kappa_2)2\gamma}.$$

The relation between $\varepsilon''(\omega)$ and γ is determined from the dispersion equation (2.29). By the expansion up to quantities of the first order in ε''

equation (2.29). By the expansion up to quantities of the first order in ε'' and γ we have

$$\gamma\left[\frac{\omega\varepsilon}{c^2}\left(\frac{1}{\kappa_1}+\frac{1}{\kappa_2}\right)+\frac{\partial\varepsilon}{\partial\omega}\left(\frac{\omega^2}{2c^2\kappa_2}-\kappa_1\right)\right]=\varepsilon''\left(\frac{\omega^2}{2c^2\kappa_2}-\kappa_1\right).$$

As a result the energy of the surface electromagnetic wave takes the form:

$$\tilde{W}=\frac{|E_0|^2}{8\pi}\frac{\omega}{4\kappa_2}\left[\frac{\omega\varepsilon}{c^2}\left(\frac{1}{\kappa_1}+\frac{1}{\kappa_2}\right)+\frac{\partial\varepsilon}{\partial\omega}\left(\frac{\omega^2}{2c^2\kappa_2}-\kappa_1\right)\right]\Big/\left(\frac{\omega^2}{2c^2\kappa_2}-\kappa_1\right).$$

This cumbersome expression becomes rather simple for the short-wavelength oscillations. In this limit

$$\tilde{W}\approx\frac{|E_0|^2}{8\pi}\frac{\omega_0}{4k}\left(\frac{\partial\varepsilon}{\partial\omega}\right)_{\omega_0}$$

where ω_0 is the frequency determined by the condition that $\varepsilon(\omega_0)=-1$.

Chapter 3

Natural Optical Activity. The Faraday and Kerr Effects

Recommended textbooks:
Landau L D and Lifshitz E M 1984 *Course of Theoretical Physics*, vol 8, *Electrodynamics of Continuous Media* (Oxford: Pergamon)
Born M and Wolf E *Principles of Optics* (Oxford: Pergamon)

If a transparent isotropic medium is not invariant with respect to reflection (inversion) about any point, that is has its stereoisomeric modification, the expansion of its dielectric tensor that takes into account a weak spatial dispersion, is as follows:

$$\varepsilon_{\alpha\beta} = \varepsilon(\omega)\delta_{\alpha\beta} + \mathrm{i}f(\omega)e_{\alpha\beta\gamma}k_\gamma \tag{3.1}$$

where $\boldsymbol{k}$ is the wave vector, and the constants ε and f are real. Such a medium has the property of natural optical activity, and the angle of rotation of the polarization plane of the electromagnetic wave is equal to (per unit path length of the ray)

$$\mathrm{d}\varphi/\mathrm{d}l = \omega^2 f/2c^2. \tag{3.2}$$

The dielectric permeability tensor of an isotropic transparent medium without spatial dispersion immersed in a weak magnetic field with the induction $\boldsymbol{B}_0$ takes the following form:

$$\varepsilon_{\alpha\beta} = \varepsilon(\omega)\delta_{\alpha\beta} + \mathrm{i}b(\omega)e_{\alpha\beta\gamma}B_{0\gamma} \tag{3.3}$$

(where ε and b are real constants). The polarization plane of an electromagnetic wave propagating in such a medium rotates by the following rule (Faraday effect):

$$\mathrm{d}\varphi/\mathrm{d}l = (b\omega B_0 \cos\theta)/2c\sqrt{\varepsilon} \tag{3.4}$$

where θ is the angle between the external magnetic field and the wave vector $\boldsymbol{k}$.

The dielectric tensor of the same medium but immersed in a weak external electric field $\boldsymbol{E}_0$ changes as follows:

$$\varepsilon_{\alpha\beta} = \varepsilon\delta_{\alpha\beta} + a_1 E_0^2\delta_{\alpha\beta} + a_2 E_{0\alpha}E_{0\beta} \tag{3.5}$$

so this medium becomes analogous to a uniaxial crystal with the optical axis directed along $\boldsymbol{E}_0$ and with the permittivities

$$\varepsilon_{\parallel} = \varepsilon + (a_1 + a_2)E_0^2 \qquad \varepsilon_{\perp} = \varepsilon + a_1 E_0^2. \tag{3.6}$$

This is called the Kerr effect.

Problem 20

Express the quantity $f(\omega)$ that arises in (3.1) and describes the natural optical activity of a medium in terms of the tensor $\varepsilon_{\alpha\beta}(\boldsymbol{r}, \boldsymbol{r}', t, t')$ that links the vectors $\mathscr{D}(\boldsymbol{r}, t)$ and $\boldsymbol{E}(\boldsymbol{r}, t)$ in formula (2.4).

For a stationary homogeneous isotropic medium, which is not invariant to inversion, the most general structure for the dielectric tensor is as follows:

$$\begin{aligned}\varepsilon_{\alpha\beta}(\boldsymbol{r}, \boldsymbol{r}', t, t') &= \varepsilon_{\alpha\beta}(\boldsymbol{\rho}, \tau) \\ &= a_1(\rho, \tau)\delta_{\alpha\beta} + a_2(\rho, \tau)\rho_\alpha\rho_\beta + a_3(\rho, \tau)e_{\alpha\beta\gamma}\rho_\gamma \qquad (3.7)\\ \rho &= \boldsymbol{r} - \boldsymbol{r}' \qquad \tau = t - t'\end{aligned}$$

since in an isotropic medium the only possible construction elements are the vector $\boldsymbol{\rho}$ and the invariant tensors $\delta_{\alpha\beta}$ and $e_{\alpha\beta\gamma}$. Because the latter is a pseudotensor the constant a_3 has to be a pseudoscalar, so it is non-zero only for a medium that is not invariant under inversion. By putting (3.7) into the formula (2.7) we obtain that

$$\begin{aligned}\varepsilon_{\alpha\beta}(\boldsymbol{k}, \omega) = \delta_{\alpha\beta}\left[1 + \int a_1\, \mathrm{e}^{-\mathrm{i}(\boldsymbol{k}\cdot\boldsymbol{\rho} - \omega t)}\, \mathrm{d}\boldsymbol{\rho}\, \mathrm{d}\tau\right] + \int a_2\rho_\alpha\rho_\beta\, \mathrm{e}^{-\mathrm{i}(\boldsymbol{k}\cdot\boldsymbol{\rho} - \omega t)}\, \mathrm{d}\boldsymbol{\rho}\, \mathrm{d}\tau \\ + e_{\alpha\beta\gamma}\int a_3\rho_\gamma\, \mathrm{e}^{-\mathrm{i}(\boldsymbol{k}\cdot\boldsymbol{\rho} - \omega t)}\, \mathrm{d}\boldsymbol{\rho}\, \mathrm{d}\tau. \qquad (3.8)\end{aligned}$$

The functions $a_1(\rho, \tau)$, $a_2(\rho, \tau)$ and $a_3(\rho, \tau)$ that enter this expression differ appreciably from zero only at distances $\rho \lesssim r_0$, where r_0 is of the order of the molecular size for an ordinary medium. At the same time the macroscopic approach itself makes sense only if r_0 is much smaller than λ—the characteristic length of the mean field variation in space. Since $\lambda = 2\pi/k$, this means that the quantity $\boldsymbol{k}\cdot\boldsymbol{\rho}$ is small ($\boldsymbol{k}\cdot\boldsymbol{\rho} \ll 1$) throughout the whole interval that gives the main contribution to the integrals in (3.8). In other words, this means that the spatial dispersion effects are usually weak in the macroscopic electrodynamics.

Thus the following expansion is valid:

$$\mathrm{e}^{-\mathrm{i}\boldsymbol{k}\cdot\boldsymbol{\rho}} \simeq 1 - \mathrm{i}\boldsymbol{k}\cdot\boldsymbol{\rho} + \cdots.$$

By substitution of the first two terms of this series into (3.8) and integration over the angles ($d\boldsymbol{\rho} = \rho^2\, d\rho\, d\Omega$) we obtain:

$$\varepsilon_{\alpha\beta}(\boldsymbol{k}, \omega) \simeq \delta_{\alpha\beta}\Bigg(1 + 4\pi \int_0^\infty e^{i\omega\tau}\, d\tau \int_0^\infty \rho^2\, d\rho\, a_1(\rho, \tau)$$

$$+ \frac{4\pi}{3} \int_0^\infty e^{i\omega\tau}\, d\tau \int_0^\infty \rho^4\, d\rho\, a_2(\rho, \tau)\Bigg)$$

$$- i e_{\alpha\beta\gamma} k_\gamma \left(\frac{4\pi}{3} \int_0^\infty e^{i\omega\tau}\, d\tau \int_0^\infty \rho^4\, d\rho\, a_3(\rho, \tau)\right). \tag{3.9}$$

The following results have been used here:

$$\int \rho_\alpha\, d\Omega = 0 \qquad \int \rho_\alpha \rho_\beta \rho_\gamma\, d\Omega = 0 \qquad \int \rho_\alpha \rho_\beta\, d\Omega = \frac{4\pi}{3} \rho^2 \delta_{\alpha\beta}.$$

Thus the expression (3.9) for the dielectric tensor $\varepsilon_{\alpha\beta}(\boldsymbol{k}, \omega)$ really has the tensor structure of the type mentioned in (3.1) (with $\varepsilon(\omega)$ and $f(\omega)$ given by the expressions in the large round brackets in (3.9)) and represents the zeroth- and the first-order terms in the expansion of the dielectric tensor into a series in the small parameter r_0/λ.

Problem 21

Find the variation of the polarization of an electromagnetic wave propagating in a medium with the dielectric tensor (3.1).

We shall start from the equations (2.9) for $\boldsymbol{E}$:

$$L_{\alpha\beta} E_\beta = \left(k_\alpha k_\beta - k^2 \delta_{\alpha\beta} + \frac{\omega^2}{c^2} \varepsilon_{\alpha\beta}\right) E_\beta = 0.$$

Let a wave propagate along the z-axis: $\boldsymbol{k} = (0, 0, k)$. Then these equations take the form:

$$\left(-k^2 + \frac{\omega^2}{c^2} \varepsilon\right) E_x + i \frac{\omega^2}{c^2} f k E_y = 0 \tag{3.10a}$$

$$\left(-k^2 + \frac{\omega^2}{c^2} \varepsilon\right) E_y - i \frac{\omega^2}{c^2} f k E_x = 0 \tag{3.10b}$$

$$(\omega^2/c^2) \varepsilon E_z = 0. \tag{3.10c}$$

For the transverse electromagnetic wave $E_z = 0$, and from equations (3.10a, b)

we obtain the following dispersion relation:

$$\left(k^2 - \frac{\omega^2}{c^2}\varepsilon\right)^2 = \frac{\omega^4}{c^4} f^2 k^2 \tag{3.11a}$$

$$k^2 - \frac{\omega^2}{c^2}\varepsilon = \pm\frac{\omega^2}{c^2} fk. \tag{3.11b}$$

The two solutions that differ by sign in (3.11*b*) correspond to the polarizations $E_x = \pm \mathrm{i} E_y$, so these are left-hand and right-hand circularly polarized electromagnetic waves. For a given frequency these waves have slightly different wave vectors. As follows from (3.11)

$$k_\pm \approx \frac{\omega}{c}\sqrt{\varepsilon} \pm \Delta k \qquad \Delta k = \frac{\omega^2}{2c^2} f \ll \frac{\omega}{c}\sqrt{\varepsilon}.$$

This difference leads to the rotation of the polarization plane of a wave. Let the electric field be directed along the x-axis at $z = 0$. By the representation of this wave as the superposition of two circularly polarized (eigensolutions), we obtain for the electric field evolution along z:

$$\boldsymbol{E} = E_0 \exp(-\mathrm{i}\omega t)[\boldsymbol{e}_+ \exp(\mathrm{i}k_+ z) + \boldsymbol{e}_- \exp(\mathrm{i}k_- z)]/\sqrt{2}$$

$$\boldsymbol{e}_\pm = (\boldsymbol{e}_x \pm \boldsymbol{e}_y)/\sqrt{2}.$$

Thus, its components

$$\begin{aligned} E_x &= \tfrac{1}{2}E_0 \exp[\mathrm{i}(\omega c^{-1}\sqrt{\varepsilon} z - \omega t)][\exp(\mathrm{i}\Delta k\, z) + \exp(-\mathrm{i}\Delta k\, z)] \\ &= E_0 \cos(\Delta k\, z) \exp[\mathrm{i}(\omega c^{-1}\sqrt{\varepsilon} z - \omega t)] \\ E_y &= \tfrac{1}{2}\mathrm{i}E_0 \exp[\mathrm{i}(\omega c^{-1}\sqrt{\varepsilon} z - \omega t)][\exp(\mathrm{i}\Delta k\, z) - \exp(-\mathrm{i}\Delta k\, z)] \\ &= -E_0 \sin(\Delta k\, z) \exp[\mathrm{i}(\omega c^{-1}\sqrt{\varepsilon} z - \omega t)] \end{aligned}$$

Since the ratio is $E_y/E_x = -\tan(\Delta k\, z)$, the wave remains linearly polarized, but for a given value of z the polarization plane is rotated through the angle $\varphi = \Delta k\, z$. Thus, the weak spatial dispersion results here in the rotation of the polarization plane:

$$\mathrm{d}\varphi/\mathrm{d}l = \Delta k = \omega^2 f/2c^2. \tag{3.12}$$

Problem 22

The molecules of a rarefied gas are chaotically oriented one spiral right-handed 'double helices'—two rigid and uniformly charged (with a specific charge of $\pm\rho$ per unit length for each) helical threads (figure 3.1). Under the action of the electric field $\boldsymbol{E}$ these threads become displaced relative to each other along their helical

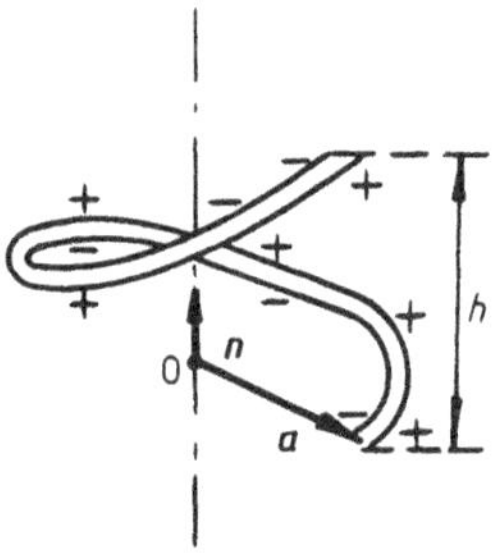

Figure 3.1. The 'double-helix' molecule.

line by the distance Δ, which is proportional to the net projection of the field on this line: $\Delta = \beta \int_0^L \boldsymbol{E}\cdot\mathrm{d}\boldsymbol{l}$. Show that such a gas has the property of natural optical activity and calculate the angle through which the polarization plane of the electromagnetic wave will be rotated (per unit length of the ray), considering that the dimension of a molecule is small in comparison with the wavelength ($a,h \ll \lambda$).

As follows from the previous problems we need to calculate the dielectric permeability tensor of such a gas with an accuracy up to the first-order terms in the small quantities a/λ and h/λ. Thus, while considering the response of a molecule (which is determined by the integral $\int_0^L \boldsymbol{E}\cdot\mathrm{d}\boldsymbol{l}$) to the electric field of the electromagnetic wave with a given frequency ω and wave vector $\boldsymbol{k}$, it is possible to write the latter in the approximate form:

$$\boldsymbol{E}(\boldsymbol{r},t) = \boldsymbol{E}_0\, \mathrm{e}^{\mathrm{i}(\boldsymbol{k}\cdot\boldsymbol{r}-\omega t)} \approx \boldsymbol{E}_0\, \mathrm{e}^{-\mathrm{i}\omega t}(1+\mathrm{i}\boldsymbol{k}\cdot\boldsymbol{r}) \tag{3.13}$$

where $\boldsymbol{E}_0$ is the field at the origin of the coordinate system (point O in figure 3.1). The orientation of a molecule is completely determined by two vectors: the unit vector $\boldsymbol{n}$ directed along the axis of the molecule's helical line, and the vector $\boldsymbol{a}$, which denotes the origin of the molecule in the plane perpendicular to $\boldsymbol{n}$ (figure 3.1). Then every point of a molecule may be determined by one parameter—the azimuthal angle φ:

$$\boldsymbol{R}(\varphi) = \boldsymbol{n}\frac{h}{2\pi}\varphi + \boldsymbol{a}\cos\varphi + (\boldsymbol{n}\times\boldsymbol{a})\sin\varphi \tag{3.14}$$

for $0 \leqslant \varphi \leqslant 2\pi$. It follows now from equations (3.13) and (3.14) that

$$\int_0^L \boldsymbol{E}\cdot\mathrm{d}\boldsymbol{l} \approx \boldsymbol{E}_0\,\mathrm{e}^{-\mathrm{i}\omega t}\int_0^{2\pi}\mathrm{d}\varphi\left[1+\mathrm{i}\left((\boldsymbol{k}\cdot\boldsymbol{n})h\frac{\varphi}{2\pi}+(\boldsymbol{k}\cdot\boldsymbol{a})\cos\varphi+\boldsymbol{k}\cdot(\boldsymbol{n}\times\boldsymbol{a})\sin\varphi\right)\right]$$
$$\times\left(\boldsymbol{n}\frac{h}{2\pi}-\boldsymbol{a}\sin\varphi+(\boldsymbol{n}\times\boldsymbol{a})\cos\varphi\right)$$

so the displacement of the threads is

$$\Delta(\boldsymbol{n}, \boldsymbol{a}, \boldsymbol{E}_0) = \beta\, \mathrm{e}^{-\mathrm{i}\omega t}\{h(\boldsymbol{E}_0\cdot\boldsymbol{n}) + \mathrm{i}(\boldsymbol{E}_0\cdot\boldsymbol{n})(\boldsymbol{k}\cdot\boldsymbol{n})h^2/2 + \mathrm{i}(\boldsymbol{E}_0\cdot\boldsymbol{a})(\boldsymbol{k}\cdot\boldsymbol{n})h - \mathrm{i}\pi(\boldsymbol{E}_0\cdot\boldsymbol{a})[\boldsymbol{k}\cdot(\boldsymbol{n}\times\boldsymbol{a})] + \mathrm{i}\pi(\boldsymbol{k}\cdot\boldsymbol{a})[\boldsymbol{E}_0\cdot(\boldsymbol{n}\times\boldsymbol{a})]\}. \quad (3.15)$$

As a result the electric charges $\pm q = \pm\rho\Delta$ appear at the origin and at the end of the molecule as well as the linear electric current $\mathscr{I} = \rho\dot{\Delta} = -\mathrm{i}\omega\rho\Delta$ along the line of the helix. Thus a molecule gains the electric ($\boldsymbol{d}$) and magnetic ($\boldsymbol{\mu}$) dipole moments

$$\boldsymbol{d} = \boldsymbol{n}hq = \rho h\Delta\boldsymbol{n}$$
$$\boldsymbol{\mu} = \frac{\mathscr{I}}{c}\pi a^2\boldsymbol{n} = \frac{-\mathrm{i}\omega\rho\Delta}{c}\pi a^2\boldsymbol{n}. \quad (3.16)$$

After this we need to calculate the specific dielectric polarization $\boldsymbol{P}$ and the magnetization $\boldsymbol{M}$ of this gas:

$$\boldsymbol{P} = n_0\langle\boldsymbol{d}\rangle \qquad \boldsymbol{M} = n_0\langle\boldsymbol{\mu}\rangle \quad (3.17)$$

where n_0 is the number of molecules per unit volume and the sign $\langle\ \rangle$ means that the quantities (3.16) are averaged over all possible orientations of the molecule. Then the density of the electric current in the gas is

$$\boldsymbol{j} = \partial\boldsymbol{P}/\partial t + c\,\mathrm{curl}\,\boldsymbol{M} = -\mathrm{i}\omega\boldsymbol{P} + \mathrm{i}c(\boldsymbol{k}\times\boldsymbol{M}) \quad (3.18)$$

so we can find the conductivity tensor $\sigma_{\alpha\beta}$ for such a gas and, hence, its dielectric permeability $\varepsilon_{\alpha\beta}$.

It is convenient to proceed with the averaging in two steps. First we consider the vector $\boldsymbol{n}$ as fixed and average over all possible orientations of the vector $\boldsymbol{a}$ in the plane orthogonal to $\boldsymbol{n}$. This may be done using the tensor notation in a way similar to that used in problem 1. For example, after substitution of (3.15) into (3.16) for $\boldsymbol{d}$, we have from the last term in (3.15) the following contribution:

$$\begin{aligned}\mathrm{i}\pi\langle\boldsymbol{n}(\boldsymbol{k}\cdot\boldsymbol{a})[\boldsymbol{E}_0\cdot(\boldsymbol{n}\times\boldsymbol{a})]\rangle_{\mathrm{a}} &= \mathrm{i}\pi\langle n_ik_la_lE_{0m}e_{mpq}n_pa_q\rangle_{\mathrm{a}} = \mathrm{i}\pi n_ik_lE_{0m}e_{mpq}n_p\langle a_la_q\rangle\\ &= \mathrm{i}\pi n_ik_lE_{0m}e_{mpq}n_p(a^2/2)(\delta_{lq} - n_ln_q)\\ &= \tfrac{1}{2}\mathrm{i}\pi a^2k_lE_{0m}e_{mpl}n_in_p. \end{aligned} \quad (3.19)$$

The fourth term in the right-hand side of (3.15) gives the same contribution while the third one becomes zero after this averaging. Making now the final averaging procedure over the vector $\boldsymbol{n}$, whose direction is randomly distributed, and taking into account the fact that

$$\langle n_in_k\rangle = \tfrac{1}{3}\delta_{ik} \qquad \langle n_in_kn_l\rangle = 0$$

we obtain the following result for the dielectric polarization of a gas $\boldsymbol{P}$:

$$\boldsymbol{P} = n_0\rho h\beta\,\mathrm{e}^{-\mathrm{i}\omega t}[\tfrac{1}{3}h\boldsymbol{E}_0 + \tfrac{1}{3}\mathrm{i}\pi a^2(\boldsymbol{k}\times\boldsymbol{E}_0)]. \quad (3.20)$$

Since, according to (3.18), the contribution to the electric current from the magnetization of the molecules is still proportional to the wave vector $\boldsymbol{k}$, we may calculate the vector $\boldsymbol{M}$ in the zeroth-order approximation, omitting the last three terms in (3.15). Thus, after averaging over $\boldsymbol{n}$ we obtain $\boldsymbol{M}$ in the form:

$$\boldsymbol{M} = \frac{-\mathrm{i}\omega n_0 \beta\rho\pi a^2 h}{3c} \boldsymbol{E}_0 \,\mathrm{e}^{-\mathrm{i}\omega t}. \tag{3.21}$$

As a result we obtain using the equations (3.18), (3.20) and (3.21) the following expression for the macroscopic current:

$$\boldsymbol{j} = \frac{-\mathrm{i}\omega n_0 \beta\rho}{3}[h^2 \boldsymbol{E} + 2\pi \mathrm{i} h a^2 (\boldsymbol{k} \times \boldsymbol{E})]$$

so the tensor $\sigma_{\alpha\beta}$ of the electric conductivity of such a gas is equal to

$$\sigma_{\alpha\beta}(k, \omega) = \frac{-\mathrm{i}h^2}{3}\omega n_0 \beta\rho\delta_{\alpha\beta} - \frac{2\pi\omega n_0 \beta\rho h a^2}{3} e_{\alpha\beta\gamma}\mathrm{k}_\gamma$$

and its dielectric permeability is

$$\varepsilon_{\alpha\beta} = \delta_{\alpha\beta} + \frac{4\pi\mathrm{i}}{\omega}\sigma_{\alpha\beta} = \varepsilon\delta_{\alpha\beta} + \mathrm{i} f e_{\alpha\beta\gamma}\mathrm{k}_\gamma \tag{3.22}$$

where

$$\varepsilon = 1 + \frac{4\pi}{3} h^2 n_0 \beta\rho \qquad f = \frac{8\pi^2}{3} n_0 \beta\rho h a^2.$$

Thus, such a gas of 'double helices' indeed possesses the natural optical activity determined by the constant f in (3.22). It is interesting to note that both the polarization and magnetization of the molecules give equal contributions to f. This is not chance but is a universal rule.

It is worth mentioning that the constant f becomes zero for $a = 0$ or $h = 0$. This is not a surprise because in these limiting cases a helical molecule is transformed into a stick or a ring, and so becomes identical to its stereoisomeric modification.

Let us show how the same results may be obtained straightforwardly from the microscopic electric currents flowing along the helical line (without introducing the dielectric polarization and magnetization of the molecules). To avoid the singularity in this calculation we introduce the finite but very small cross sectional area of a molecule, S. Then the microscopic current density in a molecule is

$$\boldsymbol{j}^{(\mathrm{m})} = \rho\dot{\Delta}\boldsymbol{t}/S = -\mathrm{i}\omega\rho\Delta(\boldsymbol{n}, \boldsymbol{a}, \boldsymbol{E}_0)\cdot\boldsymbol{t}/S$$

where $\boldsymbol{t}$ is the unit vector along the line of the molecule, equal, according to

(3.14), to

$$\boldsymbol{t} = \left[\boldsymbol{n} \frac{h}{2\pi} - \boldsymbol{a} \sin \varphi + (\boldsymbol{n} \times \boldsymbol{a}) \cos \varphi \right] \Big/ (a^2 + h^2/4\pi^2)^{1/2}.$$

In the expression given above the microscopic current at a point $\boldsymbol{r}$ is determined by the value of $\boldsymbol{E}_0$—the electric field at the origin of the coordinate system. However, to obtain the conductivity tensor $\sigma_{\alpha\beta}(\boldsymbol{k}, \omega)$ we need to know the relation between $\boldsymbol{j}$ and $\boldsymbol{E}$ at the same position in space. Thus, with the accuracy that we need it follows from (3.13) for $\boldsymbol{E}_0$ that:

$$\boldsymbol{E}_0 = \boldsymbol{E}\, \mathrm{e}^{-\mathrm{i}\boldsymbol{k}\cdot\boldsymbol{r}} \approx \boldsymbol{E}(1 - \mathrm{i}\boldsymbol{k}\cdot\boldsymbol{r}).$$

Since in the expression (3.15) for Δ all the terms, except the first one, are proportional to $\boldsymbol{k}$, they are of first order in the small parameters a/λ and h/λ, and the difference between $\boldsymbol{E}_0$ and $\boldsymbol{E}$ has to be taken into account only in the first term, putting $\boldsymbol{E}_0 \approx \boldsymbol{E}$ in all the others. Then it follows from (3.14) and (3.15) that

$$\begin{aligned} \boldsymbol{j}^{(m)} = {} & \frac{\mathrm{i}\omega\rho\beta}{S(a^2 + h^2/4\pi^2)^{1/2}} \left(\boldsymbol{n} \frac{h}{2\pi} - \boldsymbol{a} \sin \varphi + (\boldsymbol{n} \times \boldsymbol{a}) \cos \varphi \right) \\ & \times \left\{ h(\boldsymbol{E}\cdot\boldsymbol{n}) - \mathrm{i}h(\boldsymbol{E}\cdot\boldsymbol{n}) \left((\boldsymbol{k}\cdot\boldsymbol{n}) h \frac{\varphi}{2\pi} + (\boldsymbol{k}\cdot\boldsymbol{a}) \cos \varphi + \boldsymbol{k}\cdot(\boldsymbol{n} \times \boldsymbol{a}) \sin \varphi \right) \right. \\ & + \mathrm{i} \frac{h^2}{2} (\boldsymbol{E}\cdot\boldsymbol{n})(\boldsymbol{k}\cdot\boldsymbol{n}) + \mathrm{i}h(\boldsymbol{E}\cdot\boldsymbol{a})(\boldsymbol{k}\cdot\boldsymbol{n}) - \mathrm{i}\pi(\boldsymbol{E}\cdot\boldsymbol{a})[\boldsymbol{k}\cdot(\boldsymbol{n} \times \boldsymbol{a})] \\ & \left. + \mathrm{i}\pi(\boldsymbol{k}\cdot\boldsymbol{a})[\boldsymbol{E}\cdot(\boldsymbol{n} \times \boldsymbol{a})] \right\}. \end{aligned}$$

Now to obtain the macroscopic conductivity tensor for a gas, which is a continuous medium, we need to average this microscopic current over a 'physically infinitesimal' volume, that is the domain that contains a large number of molecules but whose size is very small in comparison with the wavelength $\lambda = 2\pi/k$. This type of averaging may be carried out in several steps. Firstly, the tensor coefficient that relates $\boldsymbol{j}^{(m)}$ and $\boldsymbol{E}$ is averaged over φ, that is over the different points of a molecule with vectors $\boldsymbol{n}$ and $\boldsymbol{a}$ (the vectors $\boldsymbol{n}$ and $\boldsymbol{a}$ are fixed). Secondly, this tensor has to be averaged over $\boldsymbol{a}$ and $\boldsymbol{n}$, as has been done above. Since the microscopic current is not equal to zero only inside the molecules, the result of the orientation averaging must be multiplied by the fraction of the volume occupied by molecules, that is by $n_0 SL$, where $L = 2\pi(a^2 + h^2/4\pi^2)^{1/2}$ is the length of a molecule. The final result for $\sigma_{\alpha\beta}$ and $\varepsilon_{\alpha\beta}$ is the same as in (3.22) (we omit these rather simple but cumbersome calculations).

Problem 23

Show that if a homogeneous medium with the dielectric permeability tensor (3.1) becomes slightly non-homogeneous (i.e. the characteristic length l for the variation of its parameters is much larger than the molecular size r_0), the relation between the vectors $\mathscr{D}$ and $\boldsymbol{E}$ takes the following form:

$$\mathscr{D}(\boldsymbol{r},\omega) = \varepsilon(\boldsymbol{r},\omega)\boldsymbol{E}(\boldsymbol{r},\omega) + f(\boldsymbol{r},\omega)\,\mathrm{curl}\,\boldsymbol{E}(\boldsymbol{r},\omega) + \tfrac{1}{2}[\nabla f(\boldsymbol{r},\omega) \times \boldsymbol{E}(\boldsymbol{r},\omega)]. \tag{3.23}$$

Let us write the most general relation for the vectors $\mathscr{D}(\boldsymbol{r},\omega)$ and $\boldsymbol{E}(\boldsymbol{r},\omega)$ in a isotropic medium:

$$\mathscr{D}_\alpha(\boldsymbol{r}) = E_\alpha(\boldsymbol{r}) + \int \varepsilon_{\alpha\beta}(\boldsymbol{r},\boldsymbol{r}')E_\beta(\boldsymbol{r}')\,\mathrm{d}\boldsymbol{r}'$$

$$\varepsilon_{\alpha\beta}(\boldsymbol{r},\boldsymbol{r}') = a_1(\boldsymbol{r},\boldsymbol{r}')\delta_{\alpha\beta} + a_2(\boldsymbol{r},\boldsymbol{r}')(r_\alpha - r'_\alpha)(r_\beta - r'_\beta) + a_3(\boldsymbol{r},\boldsymbol{r}')(r_\gamma - r'_\gamma)e_{\alpha\beta\gamma} \tag{3.24}$$

(compare it with the formula (3.7)).

It is important here that the functions a_1, a_2 and a_3 in this expression cannot be arbitrary. As follows from a rather general thermodynamical argument the tensor $\varepsilon_{\alpha\beta}(\boldsymbol{r},\boldsymbol{r}')$ has to satisfy the following constraint: $\varepsilon_{\alpha\beta}(\boldsymbol{r},\boldsymbol{r}') = \varepsilon_{\beta\alpha}(\boldsymbol{r}',\boldsymbol{r})$. This is a consequence of the Onsager symmetry relations for the kinetic coefficients†. By introduction of the new variables $\boldsymbol{R} = (\boldsymbol{r}+\boldsymbol{r}')/2$ and $\boldsymbol{\rho} = \boldsymbol{r} - \boldsymbol{r}'$ it is seen that this constraint would be satisfied if $a_{1,2,3}(\boldsymbol{r},\boldsymbol{r}') \equiv a_{1,2,3}(\boldsymbol{R},\rho)$. For a homogeneous medium the functions $a_{1,2,3}$ have no dependence on $\boldsymbol{R}$ and we come back to the expression (3.7). In the case of a slightly non-homogeneous medium the expansion procedure may be exploited: $\boldsymbol{R} = \boldsymbol{r} - \boldsymbol{\rho}/2$, so $a_{1,2,3}(\boldsymbol{R},\rho) \approx a_{1,2,3}(\boldsymbol{r},\rho) - \frac{1}{2}\boldsymbol{\rho}\nabla a_{1,2,3}$. By also taking into account the non-homogeneity of the electric field:

$$E_\beta(\boldsymbol{r}') \approx E_\beta(\boldsymbol{r}) - \rho_\gamma\,\partial E_\beta/\partial x_\gamma$$

and substituting this expansion into (3.24) we find

$$\mathscr{D}_\alpha(\boldsymbol{r}) = E_\alpha(\boldsymbol{r})\left(1 + \int a_1(\boldsymbol{r},\rho)\,\mathrm{d}\boldsymbol{\rho} - \frac{1}{2}\int \nabla a_1\boldsymbol{\rho}\,\mathrm{d}\boldsymbol{\rho}\right) + E_\beta(\boldsymbol{r})\left(\int a_2(\boldsymbol{r},\rho)\rho_\alpha\rho_\beta\,\mathrm{d}\boldsymbol{\rho} - \frac{1}{2}\int \nabla a_2\cdot\boldsymbol{\rho}\cdot\rho_\alpha\rho_\beta\,\mathrm{d}\boldsymbol{\rho}\right)$$

† See, Landau L D and Lifshitz E M *Course of Theoretical Physics* vol 9 (Oxford: Pergamon) ch VIII.

$$+ E_\beta(\boldsymbol{r})\left(\int a_3(\boldsymbol{r}, \rho)e_{\alpha\beta\gamma}\rho_\gamma\, \mathrm{d}\boldsymbol{\rho} - \frac{1}{2}\int \nabla a_3 \cdot \boldsymbol{\rho} \cdot e_{\alpha\beta\gamma}\rho_\gamma\, \mathrm{d}\boldsymbol{\rho}\right)$$

$$- \int \mathrm{d}\boldsymbol{\rho}\, \rho_\delta \frac{\partial E_\beta}{\partial x_\delta}[a_1(\boldsymbol{r}, \rho)\delta_{\alpha\beta} + a_2(\boldsymbol{r}, \rho)\rho_\alpha\rho_\beta + a_3(\boldsymbol{r}, \rho)e_{\alpha\beta\gamma}\rho_\gamma].$$

After integration over the angles, $\mathrm{d}\boldsymbol{\rho} = \rho^2\, \mathrm{d}\rho\, \mathrm{d}\Omega$, we obtain (similarly to the result in (3.9)):

$$\mathscr{D}_\alpha(\boldsymbol{r}) = E_\alpha(\boldsymbol{r})\varepsilon(\boldsymbol{r}) - e_{\alpha\beta\gamma}\frac{\partial E_\beta}{\partial x_\gamma}\int_0^\infty \frac{4\pi}{3} a_3(\boldsymbol{r}, \rho)\rho^4\, \mathrm{d}\rho$$

$$- \frac{1}{2} e_{\alpha\beta\gamma}\mathrm{E}_\beta \frac{\partial}{\partial \mathrm{x}_\gamma}\int_0^\infty \frac{4\pi}{3} a_3(\boldsymbol{r}, \rho)\rho^4\, \mathrm{d}\rho.$$

Thus the relation for $\mathscr{D}$ and $\boldsymbol{E}$ indeed has the form of (3.23) where

$$\varepsilon(\boldsymbol{r}, \omega) = 1 + 4\pi\int_0^\infty a_1(\boldsymbol{r}, \rho, \omega)\rho^2\, \mathrm{d}\rho + \frac{4\pi}{3}\int_0^\infty a_2(\boldsymbol{r}, \rho, \omega)\rho^4\, \mathrm{d}\rho$$

$$f(\boldsymbol{r}, \omega) = \frac{4\pi}{3}\int_0^\infty a_3(\boldsymbol{r}, \rho, \omega)\rho^4\, \mathrm{d}\rho.$$

It is worth noting that the relation (3.23) plays the important role in the definition of the boundary conditions in the electrodynamics of non-homogeneous media with spatial dispersion†.

Problem 24

Determine the variation of the polarization of an electromagnetic wave that propagates in an isotropic medium without spatial dispersion immersed in an external magnetic field with the induction $\boldsymbol{B}_0$ (the Faraday effect).

In the external magnetic field the dielectric permeability tensor of such a medium takes the form (3.3):

$$\varepsilon_{\alpha\beta} = \varepsilon\delta_{\alpha\beta} + ibe_{\alpha\beta\gamma}B_{0\gamma}.$$

† See, Ginzburg V L 1989 *Applications of Electrodynamics in Theoretical Physics and Astrophysics* (New York: Gordon and Breach) ch 2.

In the coordinate frame with the z-axis along the wave vector $\boldsymbol{k}$ and the external magnetic field coplanar to the $(x-z)$ plane the equations (2.9) for the electric field components of a wave are as follows:

$$\left(-k^2+\frac{\omega^2}{c^2}\varepsilon\right)E_x+\mathrm{i}\frac{\omega^2}{c^2}bB_{0z}E_y=0 \qquad (3.25a)$$

$$\left(-k^2+\frac{\omega^2}{c^2}\varepsilon\right)E_y-\mathrm{i}\frac{\omega^2}{c^2}bB_{0z}E_x+\mathrm{i}\frac{\omega^2}{c^2}bB_{0x}E_z=0 \qquad (3.25b)$$

$$\frac{\omega^2}{c^2}\varepsilon E_z-\frac{\omega^2}{c^2}\mathrm{i}bB_{0x}E_y=0. \qquad (3.25c)$$

Let us note first that the expression (3.3) for the dielectric tensor $\varepsilon_{\alpha\beta}$ is the approximation that describes the corrections to $\varepsilon_{\alpha\beta}$ produced by the external magnetic field up to the first order of the magnetic induction B_0. Thus, it would not be correct to take into account the higher-order terms of B_0 in equations (3.25a–c). It follows (equation (3.25c)) that the electric field component E_z parallel to $\boldsymbol{k}$ is of the first order in B_0 (remember that in an anisotropic medium the electric induction vector $\mathscr{D}$ rather than the electric field $\boldsymbol{E}$ is perpendicular to $\boldsymbol{k}$). Therefore, we may neglect the last term in (3.25b) since it is of the second order in B_0. Then the solvability condition for E_x and E_y results in the following dispersion equation:

$$k^2-\frac{\omega^2}{c^2}\varepsilon=\pm b\frac{\omega^2}{c^2}B_{0z}=\pm b\frac{\omega^2}{c^2}B_0\cos\theta \qquad (3.26)$$

where θ is the angle between the vectors $\boldsymbol{k}$ and $\boldsymbol{B}_0$. The two opposite signs in (3.26) correspond to circularly polarized electromagnetic waves ($E_y=\mp\mathrm{i}E_x$) with slightly different wave vectors

$$k_\pm=\frac{\omega}{c}\sqrt{\varepsilon}\pm\Delta k \qquad \Delta k\approx\frac{b}{2\sqrt{\varepsilon}}\frac{\omega}{c}B_0\cos\theta.$$

Thus, after calculations similar to those in problem 21 we find that the external magnetic field results in the rotation of the polarization plane with a rate

$$\frac{\mathrm{d}\varphi}{\mathrm{d}l}=\Delta k=\frac{b}{2\sqrt{\varepsilon}}\frac{\omega}{c}B_0\cos\theta.$$

The important difference with the case of natural optical activity is that in the Faraday effect the rotation rate depends on the direction in which the wave propagates.

Problem 25

Find the rate of Faraday rotation for a cold-electron plasma (see problem 6).

We may use the expression (2.15) for the dielectric permeability tensor $\varepsilon_{\alpha\beta}$ for such a plasma in the external magnetic field. By considering this field to be weak we obtain from (2.15) that

$$\varepsilon_{\alpha\beta} \approx \left(1 - \frac{\omega_{\text{pe}}^2}{\omega^2}\right)\delta_{\alpha\beta} - \frac{\mathrm{i}\omega_{\text{pe}}^2\omega_B}{\omega^3} e_{\alpha\beta\gamma}h_\gamma \qquad \boldsymbol{h} = \frac{\boldsymbol{B}_0}{B_0}. \tag{3.27}$$

Thus, it follows from (3.27) and (3.3) that the Faraday constant $b = -e\omega_{\text{pe}}^2/mc\omega^3$, so the rotation rate for this substance

$$\frac{\mathrm{d}\varphi}{\mathrm{d}l} = \frac{b}{2\sqrt{\varepsilon}}\frac{\omega}{c}B_0\cos\theta = -\frac{\omega_B}{2c}\frac{\omega_{\text{pe}}^2\cos\theta}{\omega^2(1-\omega_{\text{pe}}^2/\omega^2)}.$$

When is this simplified form (3.27) for $\varepsilon_{\alpha\beta}$ valid? For an electromagnetic wave in a plasma the frequency $\omega \geqslant \omega_{\text{pe}}$ (see the dispersion equation (2.17)). Thus, it is easy to see from (2.15) that the external magnetic field may be considered as weak if $\omega_B \ll \omega_{\text{pe}}$ (the electron gyrofrequency is much smaller than the electron plasma frequency).

Problem 26

The linearly polarized electromagnetic wave is normally incident on the plane boundary of an isotropic dielectric immersed in the external electric field E_0 (see figure 3.2), so that $(\boldsymbol{k}_0\cdot\boldsymbol{E}_0) = 0$. Find the polarization of the reflected wave.

In the coordinate frame with the x-axis along $\boldsymbol{E}_0$ the dielectric tensor $\varepsilon_{\alpha\beta}$ for

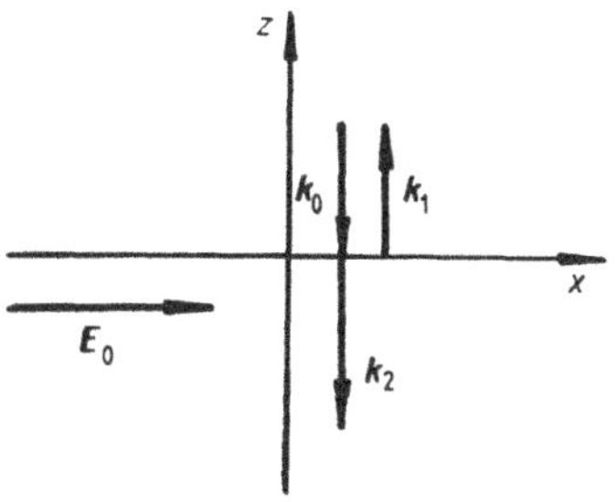

Figure 3.2. Reflection of the electromagnetic wave under the Kerr effect.

such a medium is (see (3.5)):

$$\varepsilon_{zz} = \varepsilon_{yy} = \varepsilon_{\perp} = \varepsilon + a_1 E_0^2$$

$$\varepsilon_{xx} = \varepsilon_{\parallel} = \varepsilon + (a_1 + a_2) E_0^2.$$

Let the boundary be the $z = 0$ plane, so the electric field vector of the incident wave, $\boldsymbol{E}^{(i)}$, that lies in the $(x-y)$-plane, forms the angle θ_i with $\boldsymbol{E}_0$, that is

$$E_x^{(i)} = E^{(i)} \cos\theta \qquad E_y^{(i)} = E^{(i)} \sin\theta.$$

The change in the polarization of the reflected wave is the result of the slight difference in ε_{xx} and ε_{yy} for this dielectric. Using the formula (2.26) for the case of normal incidence and considering the external electric field as sufficiently weak ($a_1 E_0^2 \sim a_2 E_0^2 \ll \varepsilon$) we obtain for the electric field of the reflected wave $E^{(r)}$:

$$\begin{aligned} E_x^{(r)} &= E_x^{(i)} \frac{\sqrt{\varepsilon_{\parallel}} - 1}{\sqrt{\varepsilon_{\parallel}} + 1} = E^{(i)} \cos\theta_i \frac{\sqrt{\varepsilon_{\parallel}} - 1}{\sqrt{\varepsilon_{\parallel}} + 1} \\ &\approx E^{(i)} \cos\theta_i \left(\frac{\sqrt{\varepsilon} - 1}{\sqrt{\varepsilon} + 1} + \frac{(a_1 + a_2) E_0^2}{2(\sqrt{\varepsilon} + 1)^2} \right) \end{aligned} \tag{3.28}$$

$$E_y^{(r)} = E_y^{(i)} \frac{\sqrt{\varepsilon_{\perp}} - 1}{\sqrt{\varepsilon_{\perp}} + 1} \approx E^{(i)} \sin\theta_i \left(\frac{\sqrt{\varepsilon} - 1}{\sqrt{\varepsilon} + 1} + \frac{a_1 E_0^2}{2(\sqrt{\varepsilon} + 1)^2} \right).$$

If this dielectric is transparent, that is if ε, a_1 and a_2 are real, the reflected wave remains linearly polarized, but with a slightly different polarization plane. It follows from (3.28) that

$$\tan\theta_r = \frac{E_y^{(r)}}{E_x^{(r)}} \approx \tan\theta_i \left(1 - \frac{a_2 E_0^2 (\sqrt{\varepsilon} - 1)}{2(\sqrt{\varepsilon} + 1)} \right)$$

so the polarization plane is rotated through the angle

$$\Delta\theta = \theta_i - \theta_r \approx \frac{a_2 E_0^2 (\sqrt{\varepsilon} - 1)}{2(\sqrt{\varepsilon} + 1)} \sin\theta \cos\theta.$$

Chapter 4

Frequency Dispersion and the Analytical Properties of the Permittivity $\varepsilon(\omega)$. The Propagation of Electromagnetic Waves in Dispersive Media

Recommended textbooks:
Landau L D and Lifshitz E M *Electrodynamics of Continuous Media* (Oxford: Pergamon)
Lipson S G and Lipson H *Optical Physics* (Cambridge: Cambridge University Press)

In an isotropic medium without spatial dispersion the relation for the electric induction $\mathscr{D}(t)$ and the electric field $\boldsymbol{E}(t)$ takes the form (see (2.4)):

$$\mathscr{D}(t) = \boldsymbol{E}(t) + \int_0^\infty f(\tau)\boldsymbol{E}(t-\tau)\,\mathrm{d}\tau \tag{4.1}$$

so the dielectric permeability tensor for such a medium is

$$\varepsilon_{\alpha\beta} = \varepsilon(\omega)\delta_{\alpha\beta}$$

where the permittivity is

$$\varepsilon(\omega) = 1 + \int_0^\infty f(\tau)\,\mathrm{e}^{\mathrm{i}\omega\tau}\,\mathrm{d}\tau. \tag{4.2}$$

The function $\varepsilon(\omega)$ is defined such that it is an *analytical function* in the upper half-plane of the complex variable ω, and $\varepsilon(\omega) \to 1$ for $|\omega| \to \infty$. The real, (ε'), and the imaginary (ε'') parts of this function are linked by *Kramers–*

Kronig relations:

$$\varepsilon'(\omega) = 1 + \frac{1}{\pi} \mathrm{P} \int_{-\infty}^{+\infty} \frac{\varepsilon''(\omega')}{\omega' - \omega} \mathrm{d}\omega'$$

$$\varepsilon''(\omega) = -\frac{1}{\pi} \mathrm{P} \int_{-\infty}^{+\infty} \frac{\varepsilon'(\omega') - 1}{\omega' - \omega} \mathrm{d}\omega'. \qquad (4.3)$$

These relations are slightly changed for a conducting medium, for which $\varepsilon(\omega)$ has a pole at $\omega = 0$ (see problem 28).

Problem 27

Find the 'memory' function of a medium $f(\tau)$ that enters formula (4.1), if its permittivity is

$$\varepsilon(\omega) = 1 - \omega_{\mathrm{p}}^2/\omega(\omega + \mathrm{i}\gamma) \qquad \gamma > 0$$

It follows from relation (4.2) that

$$\int_0^{\infty} f(\tau)\, \mathrm{e}^{\mathrm{i}\omega\tau}\, \mathrm{d}\tau = -\omega_{\mathrm{p}}^2/\omega(\omega + \mathrm{i}\gamma).$$

The function $f(\tau)$ makes physical sense only for $\tau \geqslant 0$. Let us define it formally for $\tau < 0$ by the rule

$$f(\tau) \equiv 0 \qquad \tau < 0.$$

Then we may write that

$$\int_{-\infty}^{+\infty} f(\tau)\, \mathrm{e}^{\mathrm{i}\omega\tau}\, \mathrm{d}\tau = \frac{1}{(2\pi)^{1/2}} \int_{-\infty}^{+\infty} (\sqrt{2\pi} f(\tau))\, \mathrm{e}^{\mathrm{i}\omega\tau}\, \mathrm{d}\tau = -\omega_{\mathrm{p}}^2/\omega(\omega + \mathrm{i}\gamma)$$

and consider this expression as the Fourier transform for the function $\sqrt{2\pi} f(\tau)$. Thus, the inverse transformation gives us

$$f(\tau) = -\frac{\omega_{\mathrm{p}}^2}{2\pi} \int_{-\infty}^{+\infty} \frac{\mathrm{e}^{-\mathrm{i}\omega\tau}}{\omega(\omega + \mathrm{i}\gamma)} \mathrm{d}\omega. \qquad (4.4)$$

The integrand in (4.4) has a singularity at $\omega = 0$. Thus, we need to define the calculation rules in (4.4). The answer is that we must pass above the singular point. Indeed, for $\tau < 0$ the integrand

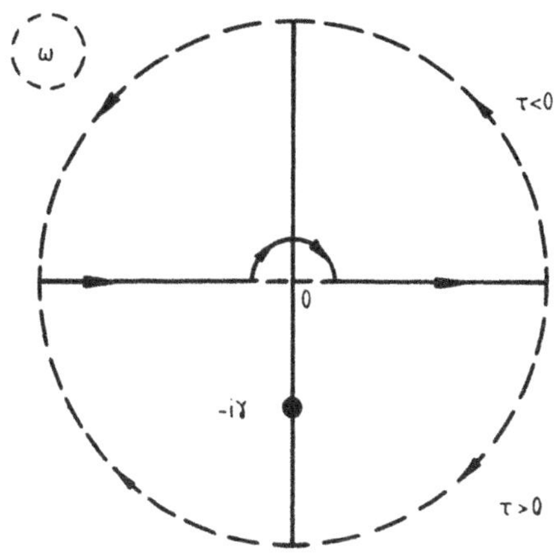

Figure 4.1. The 'passing rule' for the calculation of $f(\tau)$.

tends to zero exponentially in the upper half-plane. So for $\tau < 0$ the integral (4.4) over the real axis may be supplemented with the integral over the semicircle of sufficiently large radius in the upper half-plane (figure 4.1). Since there are no singularities inside the path of integration constructed in this way, $f(\tau) = 0$ for $\tau < 0$, as it was defined originally. To calculate $f(\tau)$ for $\tau > 0$ let us supplement (4.4) with the semicircle of sufficiently large radius in the lower half-plane. Then it is reduced to two residues at the points $\omega = -0$ *and* $\omega = -\mathrm{i}\gamma$ (taken with opposite signs):

$$f(\tau) = \frac{\omega_{\mathrm{p}}^2}{2\pi} 2\pi\mathrm{i}\left(\frac{1}{\mathrm{i}\gamma} + \frac{\mathrm{e}^{-\gamma\tau}}{(-\mathrm{i}\gamma)}\right) = \frac{\omega_{\mathrm{p}}^2}{\gamma}(1 - \mathrm{e}^{-\gamma\tau}).$$

Let us note that the 'memory' function does not tend to zero for $\tau \to +\infty$. The reason is that this medium is a conductor ($\varepsilon(\omega)$ has a pole at $\omega = 0$).

Problem 28

Determine the permittivity $\varepsilon(\omega)$ of a medium if its imaginary part is known

$$\varepsilon''(\omega) = \gamma\alpha^2/\omega(\omega^2 + \gamma^2) \qquad \gamma > 0.$$

To find the real part of permittivity, $\varepsilon'(\omega)$, we may use the Kramers–Kronig relations (4.3). Because this medium is a conductor ($\varepsilon''(\omega)$ has a pole at $\omega = 0$), the 'regularized' permittivity $\tilde{\varepsilon}(\omega)$ must be considered:

$$\tilde{\varepsilon}(\omega) = \varepsilon(\omega) - \frac{4\pi\mathrm{i}}{\omega}\sigma_0$$

where σ_0 is the static conductivity of the medium:

$$\sigma_0 = \lim_{\omega \to 0} [\omega\varepsilon''(\omega)/4\pi].$$

In our case $\sigma_0 = \alpha^2/4\pi\gamma$, so that

$$\tilde{\varepsilon}''(\omega) = \frac{\gamma\alpha^2}{\omega(\omega^2+\gamma^2)} - \frac{\alpha^2}{\omega\gamma} = -\frac{\alpha^2\omega}{\gamma(\omega^2+\gamma^2)}.$$

Then we have from the first of the relations (4.3):

$$\tilde{\varepsilon}'(\omega) = \varepsilon'(\omega) = 1 + \frac{1}{\pi}\mathrm{P}\int_{-\infty}^{+\infty} \frac{\tilde{\varepsilon}''(\omega')}{\omega'-\omega}\mathrm{d}\omega' = 1 - \frac{\alpha^2}{\pi\gamma}\mathrm{P}\int_{-\infty}^{+\infty} \frac{\omega'\mathrm{d}\omega'}{[(\omega')^2+\gamma^2](\omega'-\omega)}.$$

This integral may be easily calculated by introducing the complex-variable plane of ω' (figure 4.2). Let us supplement it with the integral along an infinitesimal semicircle around the point $\omega' = \omega$ (which is equal to the

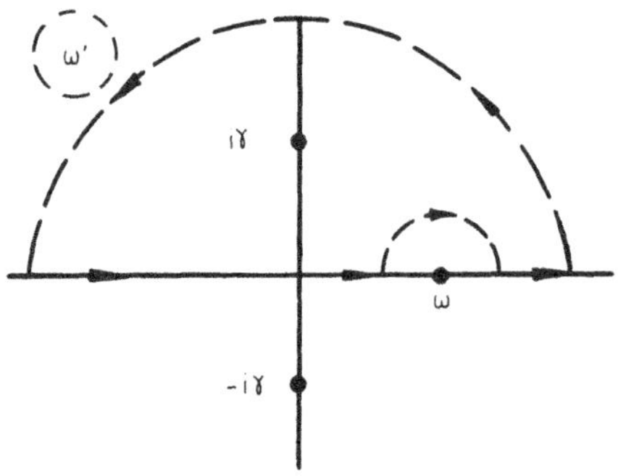

Figure 4.2. The contour of integration in the complex plane.

half-residue at this point) and with the integral over the semicircle of sufficiently large radius in the upper half-plane (which tends to zero). As a result

$$\mathrm{P}\int_{-\infty}^{+\infty} \frac{\omega'\,\mathrm{d}\omega'}{[(\omega')^2+\gamma^2](\omega'-\omega)} = 2\pi\mathrm{i}\,\mathrm{Res}|_{\omega'=\mathrm{i}\gamma} + \mathrm{i}\pi\,\mathrm{Res}|_{\omega'=\omega} = \pi\frac{\gamma}{\omega^2+\gamma^2}.$$

Thus, $\varepsilon'(\omega) = 1 - \alpha^2/(\omega^2+\gamma^2)$, and the permittivity of this medium is equal to $\varepsilon(\omega) = 1 - \alpha^2/\omega(\omega+\mathrm{i}\gamma)$.

Problem 29

Find the permittivity of a dielectric medium if the reflection coefficient $R(\omega)$ for the normal incident electromagnetic waves is measured for the whole frequency range $(0 \leqslant \omega < +\infty)$.

As is known from problem 14, for a normally incident wave the reflection coefficient is

$$R(\omega) = \left| \frac{\sqrt{\varepsilon(\omega)} - 1}{\sqrt{\varepsilon(\omega)} + 1} \right|^2 .$$

The permittivity $\varepsilon(\omega)$ may be found by a method analogous to that used in the Kramers–Kronig relations (4.3). Let us consider the complex function

$$\phi(\omega) = \frac{\sqrt{\varepsilon(\omega)} - 1}{\sqrt{\varepsilon(\omega)} + 1} \equiv \sqrt{R(\omega)}\, e^{i\varphi(\omega)}.$$

Then the function

$$\psi(\omega) \equiv \ln \phi(\omega) = \tfrac{1}{2} \ln R(\omega) + i\varphi(\omega)$$

so the problem is reduced to the determination of the imaginary part of $\psi(\omega)$ when its real part is known. Since the function $\varepsilon(\omega)$ has no singularities and null points in the upper half-plane of the complex variable ω and tends to unity only for $|\omega| \to \infty$, the functions $\phi(\omega)$ and $\psi(\omega)$ defined above are also analytical in this domain. Let us consider the integral

$$\int_\Gamma \frac{\psi(\omega)}{\omega^2 - \omega_0^2}\, d\omega$$

along the contour Γ shown in figure 4.3. Although the function $\psi(\omega)$ is logarithmically divergent at $|\omega| \to \infty$ (since in this limit $R(\omega) \to 0$), the contribution from the semicircle of sufficiently large radius is negligible

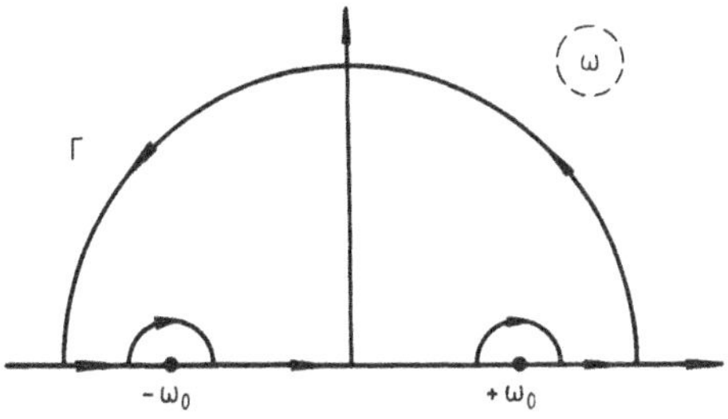

Figure 4.3. Calculation of the integral along the contour Γ.

because of the high power of the ω in the denominator in the integrand. So it follows from Cauchy's theorem that

$$\mathrm{P}\int_{-\infty}^{+\infty}\frac{\psi(\omega)\,\mathrm{d}\omega}{\omega^2-\omega_0^2}+\frac{\mathrm{i}\pi}{2\omega_0}\psi(-\omega_0)-\frac{\mathrm{i}\pi}{2\omega_0}\psi(\omega_0)=0. \tag{4.5}$$

Taking into account that the real and the imaginary parts of the function $\varepsilon(\omega)$ are the even and odd functions of ω on the real axis (see formula (4.2)), we conclude that the same is also true for the function $\psi(\omega)$. Thus, the real part of equation (4.5) gives the following relation:

$$\psi''(\omega_0)=-\frac{\omega_0}{\pi}\mathrm{P}\int_{-\infty}^{+\infty}\frac{\psi'(\omega)\,\mathrm{d}\omega}{\omega^2-\omega_0^2}$$

(the imaginary part is identically equal to zero). Then

$$\varphi(\omega)=\frac{\omega}{\pi}\mathrm{P}\int_0^{\infty}\frac{\ln R(\omega')}{\omega^2-(\omega')^2}\,\mathrm{d}\omega'$$

Now it is possible, in principle, to obtain the permittivity of a medium $\varepsilon(\omega)$.

Problem 30

Determine the propagation velocity and the shape evolution of the quasimonochromatic package of electromagnetic waves in a medium of given permittivity $\varepsilon(\omega)$.

Let the electromagnetic perturbation, in the form of a one-dimensional wave packet,

$$E(z,0)=E_0\exp(-z^2/l^2)\exp(\mathrm{i}k_0z) \tag{4.6}$$

be present in a medium at the initial moment ($t=0$). Here E is one of the transverse components of the electric field, say E_x. To determine its temporal evolution we represent (4.6) as a set of spatial harmonics:

$$E(z,0)=\frac{1}{(2\pi)^{1/2}}\int_{-\infty}^{+\infty}E(k)\,\mathrm{e}^{\mathrm{i}kz}\,\mathrm{d}k$$

where

$$E(k) = \frac{1}{(2\pi)^{1/2}} \int_{-\infty}^{+\infty} E(z,0)\,\mathrm{e}^{-\mathrm{i}kz}\,\mathrm{d}z$$

$$= \frac{E_0}{(2\pi)^{1/2}} \int_{-\infty}^{+\infty} \exp[-z^2/l^2 + \mathrm{i}(k_0 - k)z]\,\mathrm{d}z$$

$$= (E_0 l/\sqrt{2})\exp[-(k-k_0)^2 l^2/4]. \tag{4.7}$$

It is seen from (4.7) that the spectrum of this wave packet has a Gaussian shape with the characteristic width $\Delta k \sim l^{-1}$. Thus, this signal may be considered to be a quasimonochromatic one if $\Delta k \ll k_0$, that is, $k_0 l \gg 1$. Each harmonic corresponds to the frequency $\omega(k)$ given by the dispersion equation $k^2 = \omega^2 \varepsilon(\omega)/c^2$. Thus, at the subsequent moments we have

$$E(z,t) = \frac{1}{(2\pi)^{1/2}} \int_{-\infty}^{+\infty} E(k)\,\mathrm{e}^{\mathrm{i}[kz - \omega(k)t]}\,\mathrm{d}k. \tag{4.8}$$

For a quasimonochromatic packet with a narrow spectrum ($\Delta k \ll k_0$) the following expansion for $\omega(k)$ may be used:

$$\omega(k) \approx \omega(k_0) + \left.\frac{\partial\omega}{\partial k}\right|_{k_0}(k-k_0) + \frac{1}{2}\left.\frac{\partial^2\omega}{\partial k^2}\right|_{k_0}(k-k_0)^2$$

$$= \omega_0 + v_g(k-k_0) + \alpha(k-k_0)^2.$$

By substitution of this expression and the spectrum (4.7) into (4.8) we obtain that

$$E(z,t) = \phi(z,t)\exp[\mathrm{i}(k_0 z - \omega_0 t)]$$

where

$$\phi(z,t) = \frac{E_0 l}{2\sqrt{\pi}} \int_{-\infty}^{+\infty} \mathrm{d}\kappa \exp\left(-\frac{\kappa^2 l^2}{4} + \mathrm{i}\kappa(z - v_g t) - \mathrm{i}\kappa^2 \alpha t\right)$$

$$= \{\exp[-(z-v_g t)^2/(l^2 + 4\mathrm{i}\alpha t)]\}/(1 + 4\mathrm{i}\alpha t/l^2)^{1/2}.$$

The function $\phi(z,t)$ is the so-called *complex envelope* of the wave packet. The intensity of a signal is determined by the modulus of ϕ:

$$|\phi(z,t)| = \frac{l}{L(t)}\exp[-(z-v_g t)/L(t)] \qquad L(t) = (l^2 + 16\alpha^2 t^2/l^2)^{1/2} \tag{4.9}$$

It is seen now from (4.9) that a wave packet propagates with the *group velocity* $v_g = \partial\omega/\partial k$ and its width $L(t)$ is growing owing to the dispersion. For an electromagnetic wave in a dielectric medium with permittivity $\varepsilon(\omega)$

$$v_g = c\sqrt{\varepsilon}/(\varepsilon + \tfrac{1}{2}\omega\, d\varepsilon/d\omega)$$

it can be shown that v_g is less than the speed of light in vacuum, c, for any medium[†].

In the example considered above, when the initial perturbation has the form (4.6), the width of a wave packet $L(t)$ grows monotonically (see (4.9)). However, it can be shown that this is not a universal property, so it is possible to construct a quasimonochromatic signal in a dispersive medium whose width is a decreasing function of time for some interval. In the case (4.6) the wave packet has its minimal width at $t = 0$ because at this moment all the harmonics $E(k)$ have the same phases (see (4.7)). Therefore, it is possible to choose the initial phases of the harmonics in such a way that they would be equal at any given moment $t_0 > 0$, so that the width of a packet would be minimal at $t = t_0$. Put, for instance,

$$E(k) = \frac{E_0 l}{\sqrt{2}} \exp\left(-\frac{(k-k_0)^2 l^2}{4} + i\alpha(k-k_0)^2 t_0 \right)$$

then after a simple calculation similar to that made above we find an expression like (4.9) for the envelope, with the only difference now being that

$$L(t) = [l^2 + 16\alpha^2(t-t_0)^2/l^2]^{1/2}.$$

Thus the width of a packet decreases during the time interval $0 < t < t_0$, reaches its minimum at $t = t_0$ ($L_{min} = l$), and then starts to increase. It goes without saying that all these features are not in contradiction with the 'uncertainty relation'.

Problem 31

A plane electromagnetic wave packet with a sharply defined forward front is normally incident on the boundary of a half-space ($z > 0$) occupied by a dielectric of given permittivity $\varepsilon(\omega)$, so that the electric field amplitude of the incident signal at the boundary ($z = 0$) is

$$E_i(t) = \begin{cases} 0 & t < 0 \\ E_0 \sin\omega_0 t & t > 0. \end{cases}$$

Determine the electric field $E(z, t)$ inside the dielectric.

[†] See, Landau L D and Lifshitz E M 1984 *Course of Theoretical Physics* vol 8 (Oxford: Pergamon), §84.

Let us represent the incident wave packet as a number of harmonics:

$$E_i(t) = \frac{1}{(2\pi)^{1/2}} \int_{-\infty}^{+\infty} E_i(\omega)\, e^{-i\omega t}\, d\omega$$

$$E_i(\omega) = \frac{1}{(2\pi)^{1/2}} \int_{-\infty}^{+\infty} E_i(t)\, e^{i\omega t}\, dt = \frac{E_0}{(2\pi)^{1/2}} \int_0^{\infty} (\sin \omega_0 t)\, e^{i\omega t - \delta t} \tag{4.10}$$

$$= \frac{E_0}{2(2\pi)^{1/2}} \left(\frac{1}{\omega + \omega_0 + i\delta} - \frac{1}{\omega - \omega_0 + i\delta} \right).$$

We introduce here for the convenience of calculation the infinitesimal damping of the signal ($\delta > 0$), which will drop out of final results. For each of the harmonics the solution is already known (see problem 14, formula (2.26)): the amplitude of the transmitted wave is equal to $2E_i(\omega)/[1 + (\varepsilon(\omega))^{1/2}]$. Thus, the general solution for the electric field in the dielectric is

$$E(z, t) = \frac{1}{(2\pi)^{1/2}} \int_{-\infty}^{\infty} \frac{2E_i(\omega)}{1 + \varepsilon^{1/2}(\omega)} \exp\left(i \frac{\omega}{c} \varepsilon^{1/2}(\omega) z - \omega t \right) d\omega$$

$$= \frac{E_0}{2\pi} \int_{-\infty}^{+\infty} d\omega \frac{\exp\{i[\omega c^{-1} \varepsilon^{1/2}(\omega) z - \omega t]\}}{1 + \varepsilon^{1/2}(\omega)}$$

$$\times \left(\frac{1}{\omega + \omega_0 + i\delta} - \frac{1}{\omega - \omega_0 + i\delta} \right). \tag{4.11}$$

Starting from this expression let us first show that $E(z, t) = 0$ for $z > ct$. To prove this we shall consider the complex plane of a variable ω.

The field E in (4.11) is determined by the integral over the real axis. For $\delta > 0$ the poles of the integrand lie in the lower half-plane. So it is seen that the small damping included above gives the 'rule for passing' the singularities at the points $\omega = \pm\omega_0$: they must be passed from above (figure 4.4). Since the permittivity $\varepsilon(\omega) \to 1$ for $|\omega| \to \infty$, the integrand in (4.11) tends to zero exponentially at $z > ct$ for $|\omega| \to \infty$ in the upper half-plane. Therefore, we may supplement the integral (4.11) with the integral along the semicircle of sufficiently large radius there. Taking into account the analytical properties of $\varepsilon(\omega)$ in the upper half-plane we conclude that there is not any singularity inside the resulting contour of integration, so this integral is equal to zero: $E(z, t) = 0$ at $z > ct$.

What is the electric field at $z < ct$? The general expression (4.11) does not give a visible result, so we need to consider different regions where the

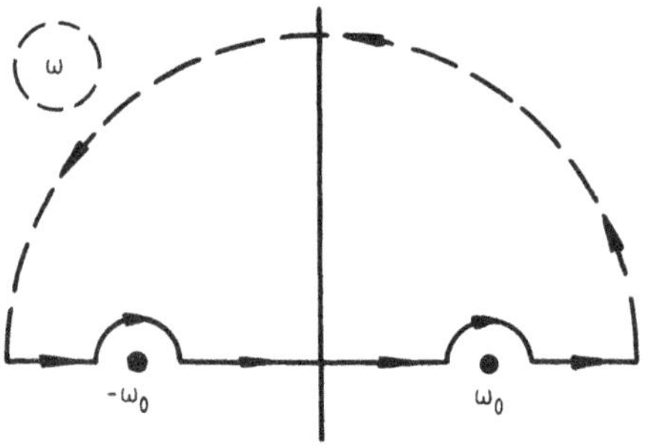

Figure 4.4. The 'passing rule' for the integration in (4.11).

contribution of some particular frequencies dominates in (4.11). As is seen from (4.10) the spectrum has a sharp maximum at the frequencies close to $\pm\omega_0$. They relate to the propagation of the 'main' signal. To simplify the calculations we may consider only one of the exponents in $\sin\omega_0 t$, so in the incident wave

$$E_i(t) = \begin{cases} 0 & t < 0 \\ E_0 \exp(-i\omega_0 t) & t > 0. \end{cases}$$

Then

$$E_i(\omega) = \frac{E_0}{(2\pi)^{1/2}} \int_0^\infty \exp[i(\omega - \omega_0)\tau]\, d\tau$$

and

$$E(z, t) = \frac{E_0}{2\pi} \int_0^{+\infty} d\tau \int_{-\infty}^{+\infty} d\omega \exp[i(\omega - \omega_0)\tau] \frac{2\exp\{i[k(\omega)z - \omega t]\}}{1 + \varepsilon^{1/2}(\omega)}. \quad (4.12)$$

Since we are interested in the contribution of the frequencies close to ω_0, the following expansion may be used in the exponent in (4.12):

$$\omega = \omega_0 + \xi$$

$$k(\omega) \approx k(\omega_0) + \left.\frac{dk}{d\omega}\right|_{\omega_0} \xi + \frac{1}{2}\left.\frac{d^2k}{d\omega^2}\right|_{\omega_0} \xi^2 = k_0 + \frac{\xi}{v_g} - \frac{\xi^2}{2}\frac{v_g'}{v_g^2}.$$

As a result

$$E(z, t) = E_0 \exp[i(k_0 z - \omega_0 t)] \frac{2}{1 + \varepsilon^{1/2}(\omega_0)} \mathscr{F}(z, t)$$

where the envelope

$$\mathscr{F}(z, t) = \frac{1}{2\pi} \int_0^\infty d\tau \int_{-\infty}^{+\infty} d\xi \exp\left[i\xi\tau + i\left(\frac{\xi}{v_g}(z - v_g t) - \frac{\xi^2}{2}\frac{v_g'}{v_g^2} z\right)\right].$$

After the simple integration over ξ we find that

$$\mathscr{F}(z, t) = \frac{1}{(i\pi)^{1/2}} \int_{y_0}^{\infty} \exp(iy^2)\, dy \qquad y_0 = \frac{(z - v_g t)}{(2z|v_g'|)^{1/2}}.$$

As may be expected in advance the front of the main signal propagates into the medium with the group velocity v_g. The width of the front grows with time because of dispersion:

$$\Delta z \sim (z|v_g'|)^{1/2} \sim (v_g|v_g'|t)^{1/2}.$$

Far enough behind a front ($y_0 < 0, |y_0| \gg 1$) the function $\mathscr{F}(z, t)$ tends to unity. So the Fresnel solution for a monochromatic wave with the frequency $\omega = \omega_0$ is established in this region.

Although the spectrum (4.10) has a maximum at $\omega = \pm\omega_0$, $E_i(\omega)$ differs from zero at any given frequency. In particular, very high frequencies are also present in the spectrum. Since the permittivity of a medium $\varepsilon(\omega)$ tends to unity for such a frequency, their propagation velocity is very close to the speed of light c. Therefore, at large distances preceding the main signal considered above the so-called 'precursor' is formed that propagates with the velocity c. To calculate the electric field in the precursor the asymptotic form for $\varepsilon(\omega)$ at high frequencies may be used: $\varepsilon(\omega) \approx 1 - \omega_{pe}^2/\omega^2$, where ω_{pe} is the electron plasma frequency for a medium. It then follows from (4.11) in the high-frequency limit that

$$E(z, t) \approx -\frac{E_0\omega_0}{2\pi} \int_{\Gamma} \frac{d\omega}{\omega^2} \exp\left[i\left(\frac{\omega}{c}(z - ct) - \frac{\omega_{pe}^2}{2\omega} t \right) \right]. \qquad (4.13)$$

The integral (4.13) has to be calculated along the path shown in figure 4.5.

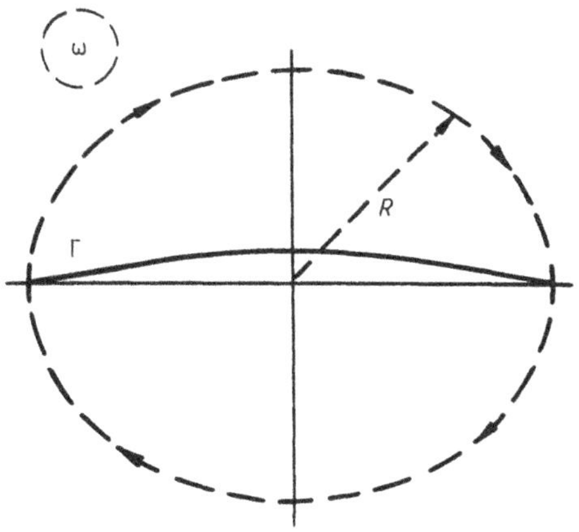

Figure 4.5. The integration contour for the precursor.

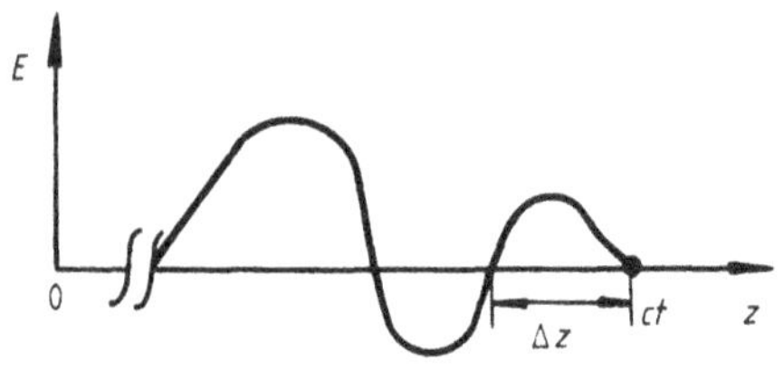

Figure 4.6. The pulse shape for the precursor.

Let us now deform this path into the semicircle of sufficiently large radius in the upper half-plane. Since for $z - ct < 0$ the integrand in (4.13) tends to zero exponentially in the lower half-plane, we may supplement (4.13) with the integral along a semicircle of sufficiently large radius in the lower half-plane. As a result the integration may be carried out along the circle of a large radius R in the plane of complex variable ω. Let us choose $R = [\omega_{pe}^2 ct/2(ct - z)]^{1/2}$. In the precursor region $\xi \equiv (ct - z)/ct \ll 1$, so this radius is really large. For $\omega = R\,e^{i\varphi}$ it follows from (4.13) that

$$E(z,t) \approx -\frac{iE_0\omega_0}{2\pi}\int_0^{2\pi}\frac{d\varphi}{R\,e^{i\varphi}}\exp\left[i\left(-\omega_{pe}t\sqrt{\frac{\xi}{2}}\,e^{i\varphi} - \omega_{pe}t\sqrt{\frac{\xi}{2}}\,e^{-i\varphi}\right)\right]$$

$$= \frac{E_0\omega_0}{\omega_{pe}}(2\xi)^{1/2}\mathscr{I}_1(\omega_{pe}t(2\xi)^{1/2})$$

$$= \frac{E_0\omega_0}{\omega_{pe}}\left(\frac{2(ct-z)}{ct}\right)^{1/2}\mathscr{I}_1\left(\omega_{pe}t\left(\frac{2(ct-z)}{ct}\right)^{1/2}\right) \tag{4.14}$$

where $\mathscr{I}_1$ is the Bessel function of the first kind with the following integral representation:

$$\mathscr{I}_1(x) = \frac{1}{2\pi}\int_{-\pi}^{\pi} e^{i(x\sin\varphi - \varphi)}\,d\varphi.$$

Taking into account the well-known properties of the Bessel functions†, we may conclude from (4.14) that the precursor is a sequence of pulses (see figure 4.6) whose width and amplitude reduce as it propagates inside the dielectric:

$$E \sim E_0\frac{\omega_0}{\omega_{pe}^2 t} \qquad \Delta z \sim \frac{c}{\omega_{pe}^2 t}.$$

† See, for instance, Abramowitz M and Stegun I A *Handbook of Mathematical Functions* (New York: Dover) §9.

Chapter 5

Cherenkov Radiation. Transition Radiation

Recommended textbooks:
Landau L D and Lifshitz E M *Electrodynamics of Continuous Media* (Oxford: Pergamon)
Jackson J D *Classical Electrodynamics* (New York: Wiley)

Problem 32

Determine the power of Cherenkov radiation for the Dirac monopole with a magnetic charge q moving with the constant velocity $\boldsymbol{v}$ in a dielectric of given permittivity, $\varepsilon(\omega)$.

In the presence of magnetic charges the Maxwell equations in a medium take the following form:

$$\begin{aligned} &\operatorname{div} \mathscr{D} = 4\pi\rho_{\mathrm{ex}}^{(\mathrm{E})} \qquad &\operatorname{div} \boldsymbol{B} = 4\pi\rho_{\mathrm{ex}}^{(\mathrm{M})} \\ &\operatorname{curl} \boldsymbol{B} = \frac{1}{c}\frac{\partial \mathscr{D}}{\partial t} + \frac{4\pi}{c}\boldsymbol{j}_{\mathrm{ex}}^{(\mathrm{E})} \qquad &\operatorname{curl} \boldsymbol{E} = -\frac{1}{c}\frac{\partial \boldsymbol{B}}{\partial t} - \frac{4\pi}{c}\boldsymbol{j}_{\mathrm{ex}}^{(\mathrm{M})} \end{aligned} \tag{5.1}$$

where $\rho_{\mathrm{ex}}^{(\mathrm{E})}, \boldsymbol{j}_{\mathrm{ex}}^{(\mathrm{E})}$ and $\rho_{\mathrm{ex}}^{(\mathrm{M})}, \boldsymbol{j}_{\mathrm{ex}}^{(\mathrm{M})}$ are the external charge densities and currents of the electric and magnetic charges. In our case $\rho_{\mathrm{ex}}^{(\mathrm{E})}$ and $\boldsymbol{j}_{\mathrm{ex}}^{(\mathrm{E})}$ are absent, and

$$\rho_{\mathrm{ex}}^{(\mathrm{M})} = q\delta(\boldsymbol{r} - \boldsymbol{v}t) \qquad \boldsymbol{j}_{\mathrm{ex}}^{(\mathrm{M})} = q\boldsymbol{v}\delta(\boldsymbol{r} - \boldsymbol{v}t).$$

After Fourier transformation we obtain from the two last equations in (5.1) that

$$\boldsymbol{B}(\boldsymbol{k}, \omega) = \frac{2iq[(\varepsilon\omega\boldsymbol{v}/c^2) - \boldsymbol{k}]}{(k^2 - \varepsilon\omega^2/c^2)}\delta(\omega - \boldsymbol{k}\cdot\boldsymbol{v})$$

so $\boldsymbol{B}(\boldsymbol{r}, t)$ is equal to

$$\boldsymbol{B}(\boldsymbol{r}, t) = \frac{1}{(2\pi)^2}\int \boldsymbol{B}(\boldsymbol{k}, \omega)\, \mathrm{e}^{\mathrm{i}(\boldsymbol{k}\cdot\boldsymbol{r} - \omega t)}\, \mathrm{d}\boldsymbol{k}\, \mathrm{d}\omega$$

$$= \frac{\mathrm{i}q}{2\pi^2}\int \mathrm{d}\boldsymbol{k}\, \frac{(\varepsilon\boldsymbol{v}/c^2)(\boldsymbol{k}\cdot\boldsymbol{v}) - \boldsymbol{k}}{[k^2 - \varepsilon(\boldsymbol{k}\cdot\boldsymbol{v})^2/c^2]}\, \mathrm{e}^{\mathrm{i}[\boldsymbol{k}\cdot\boldsymbol{r} - (\boldsymbol{k}\cdot\boldsymbol{v})t]}. \tag{5.2}$$

In a transparent medium the energy losses of a charge are only the result of radiation, so its power $\mathscr{I}$ is related to the drag force $\mathscr{F}$ acting on the charge:

$$\mathscr{I} = -\mathscr{F}\cdot\boldsymbol{v} = -q\boldsymbol{v}\cdot\boldsymbol{B}|_{r = vt}$$

(it is clear from the symmetry arguments that the electric field does not contribute to $\mathscr{F}$).

Thus, with the help of (5.2), we find that

$$\mathscr{I} = -\frac{\mathrm{i}q^2}{2\pi^2}\int \frac{\mathrm{d}\boldsymbol{k}\,(\varepsilon v^2/c^2 - 1)(\boldsymbol{k}\cdot\boldsymbol{v})}{k^2 - \varepsilon(\boldsymbol{k}\cdot\boldsymbol{v})/c^2}$$

$$= \frac{-\mathrm{i}q^2}{2\pi^2 v}\int_{-\infty}^{+\infty} \mathrm{d}\omega \int \mathrm{d}\boldsymbol{k}_\perp\, \frac{\omega[\varepsilon(\omega)v^2/c^2 - 1]}{k_\perp^2 + \omega^2/v^2 - \varepsilon(\omega)\omega^2/c^2}. \tag{5.3}$$

We introduce new variables of integration into (5.3): the component of the wave vector, $\boldsymbol{k}_\perp$, perpendicular to $\boldsymbol{v}$ and the frequency $\omega = (\boldsymbol{k}\cdot\boldsymbol{v})$, so that $\mathrm{d}\boldsymbol{k} = \mathrm{d}\boldsymbol{k}_\perp\, \mathrm{d}\omega/v$. Since the frequencies $\pm\omega$ correspond physically to the same wave (only a positive frequency makes physical sense), the spectral power of the energy losses of the monopole may be written, according to (5.3), in the form:

$$\frac{\mathrm{d}\mathscr{I}}{\mathrm{d}\omega} = -\,-\frac{\mathrm{i}q^2}{2\pi^2 v}\int \mathrm{d}\boldsymbol{k}_\perp \sum_{+,-} \frac{\omega[\varepsilon(\omega)(v^2/c^2) - 1]}{k_\perp^2 + \omega^2/v^2 - \varepsilon(\omega)\omega^2/c^2} \tag{5.4}$$

where the symbol $\sum_{+,-}$ means the sum of the contributions from the positive and negative ω. As will be seen from what follows, Cherenkov radiation is related to a possible zero of the denominator in the integrand (5.4). So for its correct calculation it is convenient to introduce a weak dissipation in a medium, thus a small imaginary part of the permittivity, $\varepsilon''(\omega)$, becomes present. Taking into account the fact that $\varepsilon''(\omega)$ is an odd function of ω, and the real part, $\varepsilon'(\omega)$, is an even one (these properties of $\varepsilon(\omega)$ follow from (4.2)), we find from (5.4) that

$$\frac{\mathrm{d}\mathscr{I}}{\mathrm{d}\omega} = \frac{-\mathrm{i}q^2}{\pi v}\int_0^\infty k_\perp\, \mathrm{d}k_\perp\, \omega\left(\frac{\varepsilon' v^2}{c^2} - 1\right)\frac{2\mathrm{i}(\omega^2/c^2)\varepsilon''}{(k_\perp^2 + \omega^2/v^2 - \varepsilon'(\omega^2/c^2))^2 + [(\omega^2/c^2)\varepsilon'']^2}.$$

Since when ε'' tends to zero

$$\lim_{\varepsilon''\to 0} \frac{(\omega^2/c^2)\varepsilon''}{(k_\perp^2 + \omega^2/v^2 - (\omega^2/c^2)\varepsilon')^2 + ((\omega^2/c^2)\varepsilon'')^2} = \pi\delta\left(k_\perp^2 + \frac{\omega^2}{v^2} - \frac{\omega^2}{c^2}\varepsilon'\right)$$

in the transparent medium

$$\frac{\mathrm{d}\mathscr{I}}{\mathrm{d}\omega} = \frac{q^2\omega}{v}\left(\frac{\varepsilon(\omega)v^2}{c^2} - 1\right)\int_0^\infty \mathrm{d}(k_\perp^2)\,\delta\left(k_\perp^2 + \frac{\omega^2}{v^2} - \frac{\omega^2}{c^2}\varepsilon(\omega)\right) \qquad (5.5)$$

It is seen now that radiation is possible only if the inequality $v > c/\varepsilon^{1/2}(\omega)$ is satisfied (otherwise the argument of the δ-function in (5.5) cannot be turned to zero). The Cherenkov cone with $\cos\theta = c/v\varepsilon^{1/2}(\omega)$ relates to any frequency ω (for which $\varepsilon^{1/2}(\omega) > c/v$), where θ is the angle between the wave vector $\boldsymbol{k}$ of a radiated electromagnetic wave and the velocity of the monopole $\boldsymbol{v}$ (this relation follows from the condition $\omega = \boldsymbol{k}\cdot\boldsymbol{v} = kv\cos\theta = (\omega/c)\varepsilon^{1/2}(\omega)v\cos\theta$). Finally, we find from (5.5) that

$$\frac{\mathrm{d}\mathscr{I}}{\mathrm{d}\omega} = \begin{cases} \dfrac{q^2\omega}{v}\left([\varepsilon(\omega)v^2/c^2] - 1\right) & \text{for } \varepsilon(\omega) > c^2/v^2 \\ 0 & \text{for } \varepsilon(\omega) < c^2/v^2. \end{cases}$$

Problem 33

Determine the power of the Cherenkov radiation for a relativistic neutron moving with a constant velocity $\boldsymbol{v}$ in a medium of given permittivity $\varepsilon(\omega)$. The magnetic moment of the neutron is directed perpendicularly to its velocity.

The neutron possesses its own dipole magnetic moment equal to m_0 in the frame of reference of the neutron (the frame where the neutron is at rest). In the laboratory frame the neutron has electric and magnetic moments both of which depend on the orientation of the neutron spin. Taking into account that the electric moment of the neutron is equal to zero, we find†

$$\boldsymbol{d} = (\boldsymbol{v}\times\boldsymbol{m}_0)/c \qquad \boldsymbol{m} = \boldsymbol{m}_0 - (\gamma - 1)\boldsymbol{v}\cdot(\boldsymbol{m}_0\cdot\boldsymbol{v})/\gamma v^2 \qquad (5.6)$$

where $\boldsymbol{d}$ and $\boldsymbol{m}$ are the electric and the magnetic moments of the neutron in the laboratory frame of reference, and γ is the neutron's relativistic factor. Since in our case $\boldsymbol{m}_0 \perp \boldsymbol{v}$, $\boldsymbol{m} = \boldsymbol{m}_0$. The external current, produced by the neutron in a medium, is

$$\boldsymbol{j}_{\mathrm{ex}} = \partial\boldsymbol{P}/\partial t + c\,\mathrm{curl}\,\boldsymbol{M}$$

† See, for instance, Ginzburg V L 1989 *Applications of Electrodynamics in Theoretical Physics and Astrophysics* (New York: Gordon and Breach).

where

$$P(r, t) = d\delta(r - vt) \qquad M(r, t) = m_0\delta(r - vt).$$

So

$$j_{ex}(k, \omega) = \frac{i\delta(\omega - k\cdot v)}{2\pi}[-\omega d + c(k \times m_0)]$$

and after Fourier transformation in the Maxwell equations we find that

$$E(k, \omega) = \left(k^2 - \frac{\omega^2\varepsilon(\omega)}{c^2}\right)^{-1}\left(\frac{\omega^2}{c^2}d - \frac{\omega}{c}(k \times m_0) - \frac{k(k\cdot d)}{\varepsilon(\omega)}\right)2\delta(\omega - k\cdot v)$$

$$(5.7)$$

$$B(k, \omega) = \left(k^2 - \frac{\omega^2\varepsilon(\omega)}{c^2}\right)^{-1}\left(\frac{\omega}{c}(k \times d) - [k \times (k \times m_0)]\right)2\delta(\omega - k\cdot v).$$

The drag force acting on the neutron is

$$\mathscr{F} = [(d\cdot\nabla)E + \nabla(m_0\cdot B)]_{r = vt}.$$

With the help of (5.7) it may be written as follows:

$$\mathscr{F} = (2\pi)^{-2}\int\frac{e^{i[(k\cdot v)t - \omega t]}}{k^2 - \omega^2\varepsilon(\omega)/c^2}dk\, d\omega\, 2\delta(\omega - k\cdot v)$$
$$\times\left[i(k\cdot d)\left(\frac{\omega^2}{c^2}d - \frac{\omega}{c}(k \times m_0) - \frac{k(k\cdot d)}{\varepsilon(\omega)}\right)\right.$$
$$\left. + ik\left(\frac{\omega}{c}m_0\cdot(k \times d) - (k\cdot m_0)^2 + k^2m_0^2\right)\right].$$

Thus the drag power $\mathscr{I} = -\mathscr{F}\cdot v$ is

$$\mathscr{I} = \frac{-i}{2\pi^2}\int\frac{dk}{\{k^2 - [(kv)^2/c^2]\varepsilon(k\cdot v)\}}\left[(k\cdot d)\left(-\frac{(k\cdot v)}{c}[v\cdot(k \times m_0)] - \frac{(k\cdot v)}{\varepsilon(k\cdot v)}(k\cdot d)\right)\right.$$
$$\left. + (k\cdot v)\left(k^2m_0^2 - (k\cdot m_0)^2 - \frac{(k\cdot v)^2m_0^2}{c^2}\right)\right]$$

$$(5.8)$$

(we use here that the vectors m_0 and d are perpendicular to v). After simple calculations analogous to those in the previous problem we find for the spectral radiation power:

$$\frac{d\mathscr{I}}{d\omega} = \frac{\omega}{\pi v}\int dk_\perp\,\delta\left(k_\perp^2 + \frac{\omega^2}{v^2} - \frac{\omega^2}{c^2}\varepsilon(\omega)\right)$$
$$\times\left[(k_\perp\cdot d)^2\left(1 - \frac{1}{\varepsilon(\omega)}\right) + [k_\perp^2m_0^2 - (k_\perp\cdot m_0)^2]\right]. \qquad (5.9)$$

Unlike the case of a point charge the azimuthal symmetry around the velocity vector of the radiation power is now absent, so the latter is dependent on the angle φ between the vectors $\boldsymbol{k}_\perp$ and $\boldsymbol{m}_0$. Let, for simplicity, the neutron to be ultrarelativistic, so that $\gamma \gg 1$. It then follows from (5.6) that $d \approx m_0$, and after the integration over the modulus of $k_\perp$ in (5.9) we find

$$\frac{\mathrm{d}\mathscr{I}}{\mathrm{d}\omega\,\mathrm{d}\varphi} = \frac{m_0^2\omega^3[\varepsilon(\omega)-1]}{2\pi c^3}[2-\varepsilon^{-1}(\omega)]\sin^2\varphi.$$

This expression gives the azimuthal distribution of the radiation power. The net power is

$$\frac{\mathrm{d}\mathscr{I}}{\mathrm{d}\omega} = \begin{cases} \dfrac{m_0^2\omega^3}{2c^3}[\varepsilon(\omega)-1][2-\varepsilon^{-1}(\omega)] & \text{for} \quad \varepsilon(\omega) \geqslant 1 \\ 0 & \text{for} \quad \varepsilon(\omega) < 1. \end{cases}$$

Problem 34

Find the amplitude of the Langmuir oscillations (see problem 7) excited in a cold-electron plasma by the homogeneously charged plane surface moving with the constant velocity u in the normal (to this plane) direction.

Let the plane have the surface charge density σ and move along the z-axis. Then the electric field $E_z \equiv E$, the charge density in the plasma ρ and the electric current $j \equiv j_z$ are dependent only on z and t, and satisfy the following equations:

$$\partial E/\partial z = 4\pi\rho + 4\pi\rho_{\mathrm{ex}} \qquad \partial\rho/\partial t + \partial j/\partial z = 0$$
$$\rho_{\mathrm{ex}} = \sigma\delta(z - ut). \tag{5.10}$$

By differentiation of the second of these with respect to time:

$$\frac{\partial^2\rho}{\partial t^2} + \frac{\partial^2 j}{\partial z\,\partial t} = 0$$

and taking into account that in the linear approximation

$$j = nev \qquad \partial j/\partial t = ne\,\partial v/\partial t = ne^2E/m = \omega_{\mathrm{pe}}^2(E/4\pi)$$

(n, e, m and v are the electron's density, charge, mass and velocity, respectively) we find that

$$\frac{\partial^2\rho}{\partial t^2} + \frac{\omega_{\mathrm{pe}}^2}{4\pi}\frac{\partial E}{\partial z} = 0.$$

Together with the first of the equations (5.10) it gives

$$\partial^2\rho/\partial t^2 + \omega_{\mathrm{pe}}^2\rho = -\omega_{\mathrm{pe}}^2\rho_{\mathrm{ex}} = -\sigma\omega_{\mathrm{pe}}^2\delta(z-ut). \tag{5.11}$$

This equation is the same as for an oscillator excited by the external force which represents an instantaneous push at the moment $t = z/u$. Its solution may be obtained by integration of (5.11) over the infinitesimal interval around this moment, and gives

$$\left.\frac{\partial \rho}{\partial t}\right|_{z/u+\varepsilon} = -\frac{\sigma\omega_{pe}^2}{u}.$$

Solving equation (5.11) for the free oscillator with this initial condition (and taking into account the fact that $\rho|_{z/u} = 0$) we find the charge density perturbation in a plasma:

$$\rho(z,t) = \begin{cases} -(\sigma\omega_{pe}/u)\sin\omega_{pe}(t - z/u) & t > z/u \\ 0 & t < z/u. \end{cases} \tag{5.12}$$

Thus a wake is formed behind the moving charged plane with perturbations of charge density and electric field (Langmuir oscillations). From (5.12) and the Poisson equation we obtain for the electric field of the wake:

$$E(z,t) = \begin{cases} -4\pi\sigma\cos\omega_{pe}(t - z/u) & t > z/u \\ 0 & t < z/u. \end{cases} \tag{5.13}$$

It is seen from this expression that the wave vector of the excited Langmuir oscillations is equal to ω_{pe}/u, so the usual Cherenkov resonance condition $\omega = ku$ is satisfied.

Let us consider now the energy balance for this process. Since Langmuir waves are continuously excited, their total energy (per unit area of the moving plane) grows with the rate

$$\frac{d\mathscr{E}}{dt} = Wu = \frac{E_0^2}{8\pi}u = \frac{(4\pi\sigma)^2}{8\pi}u = 2\pi\sigma^2 u$$

where W is the energy per unit volume for this wave (see problem 17). Hence the drag force (per unit area) acting on the moving charged plane must be equal to $\mathscr{F} = u^{-1}\,d\mathscr{E}/dt = 2\pi\sigma^2$. This is indeed the case here. This drag force is actually equal to $-\sigma E_m$, where E_m is the electric field produced by the medium itself at the position of the moving plane. Since the electric field of the charged plane is

$$E_p = \begin{cases} -2\pi\sigma & z < ut \\ 2\pi\sigma & z > ut \end{cases}$$

it follows from (5.13) that

$$E_m(z = ut) = E(z = ut) - E_p = -2\pi\sigma$$

so the energy balance is really satisfied here.

Problem 35

A non-relativistic point charge q is moving in a vacuum with a constant velocity $\boldsymbol{v}$ normal to the plane boundary of an ideally conducting medium. Determine the spectral and the angular distributions and the polarization of the transition radiation excited by the charge while crossing this boundary.

In the case of an ideally conducting medium its polarization is equivalent to the appearance of a fictitious charge $-q$ in the symmetrical point with respect to the boundary plane (figure 5.1). The dipole electric moment of such a system of charges

$$\boldsymbol{d} = \begin{cases} 2q\boldsymbol{v}t & t \leqslant 0 \\ 0 & t > 0 \end{cases} \tag{5.14}$$

(we consider here that the charge reaches the boundary at the moment $t = 0$).

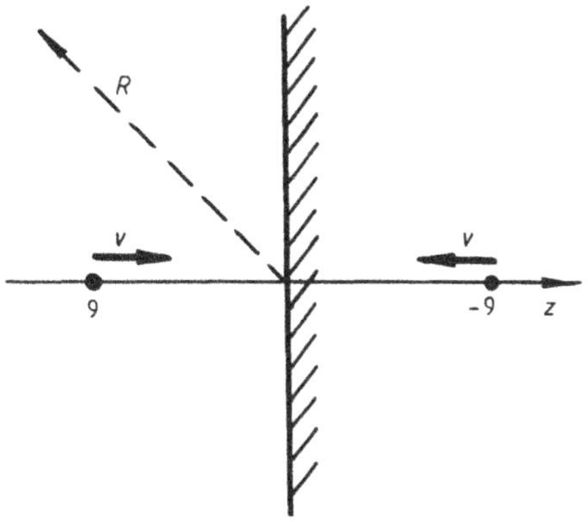

Figure 5.1. A simple model for the transition radiation.

Since the vector $\ddot{\boldsymbol{d}} = -2q\boldsymbol{v}\delta(t)$, a burst of transition radiation arises at the moment $t = 0$. According to the well known electrodynamical relations†, for the dipole radiation we have in the wave zone:

$$\boldsymbol{E} = \frac{1}{c^2 R}[(\ddot{\boldsymbol{d}} \times \boldsymbol{n}) \times \boldsymbol{n}] \qquad \boldsymbol{B} = \frac{1}{c^2 R}(\ddot{\boldsymbol{d}} \times \boldsymbol{n}) \qquad \boldsymbol{n} = \frac{\boldsymbol{R}}{R}.$$

Thus, it is seen at once that this radiation is linearly polarized with the vector $\boldsymbol{B}$

† See, for instance, Landau L D and Lifshitz E M 1971 *Course of Theoretical Physics* vol 2 (Oxford: Pergamon) ch IX.

perpendicular to the plane formed by the wave vector of the radiated wave and $\boldsymbol{v}$. For the spectral and angular distributions of the radiation we have

$$\frac{d\mathscr{E}}{d\omega\, d\Omega} = \frac{c}{4\pi}|B_\omega|^2 R^2 = \frac{[(\ddot{\boldsymbol{d}})_\omega \times \boldsymbol{n}]^2}{2\pi c^3}.$$

By calculating the Fourier transforms in (5.14) we find that $(\ddot{\boldsymbol{d}})_\omega = -2q\boldsymbol{v}/(2\pi)^{1/2}$, so that

$$\frac{d\mathscr{E}}{d\omega\, d\Omega} = q^2(\boldsymbol{v} \times \boldsymbol{n})^2/\pi^2 c^3. \qquad (5.15)$$

Thus the net radiated energy, equal to the integral of (5.15) over all frequencies and directions, becomes divergent at high frequencies (the spectrum (5.15) has no dependence on ω). The physical reason for this difficulty is that the model of an ideal conductor is not adequate at high frequencies, because $\varepsilon(\omega)$ tends to unity in this limit for any medium. Thus, the expression (5.15) obtained for this idealized model is valid only in the frequency range where $|\varepsilon(\omega)| \gg 1$. For the conductor this means that $\omega \ll \sigma_0$, where σ_0 is the static conductivity of a medium.

Problem 36

The non-relativistic point charge q moves with constant velocity $\boldsymbol{v}$ in a direction perpendicular to the plane boundary between the vacuum and the dielectric (figure 5.2). What is the energy of the surface electromagnetic waves excited by the charge while crossing this boundary?

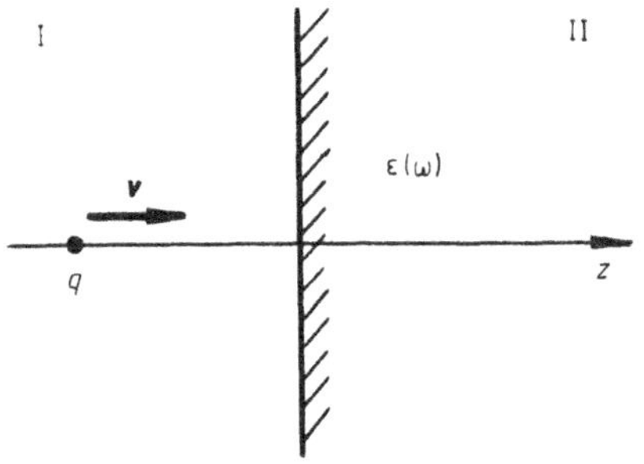

Figure 5.2. Excitation of surface waves by a moving point charge.

Since the charge is non-relativistic ($v \ll c$), it is possible to consider the electric field $\boldsymbol{E}$ created to be approximately a potential one: $\boldsymbol{E} = -\nabla\varphi$. Then

$$\operatorname{div}(\hat{\varepsilon}\boldsymbol{E}) = -\operatorname{div}(\hat{\varepsilon}\nabla\varphi) = 4\pi\rho_{ex} \qquad (5.16)$$

where $\hat{\varepsilon}$ is the operator of dielectric permeability and $\rho_{ex} = q\delta(\boldsymbol{r} - \boldsymbol{v}t)$. Let us note the boundary as the plane $z = 0$. Then after a Fourier transformation on the transverse coordinates (x, y) and time we solve equation (5.16) for the Fourier transform of the potential $\varphi(\boldsymbol{k}_\perp, \omega, z)$ separately in a vacuum ($z < 0$, domain I) and in a dielectric ($z > 0$, domain II). Taking into account the fact that

$$\rho_{ex}(\boldsymbol{k}_\perp, \omega, z) = (2\pi)^{-3/2} \int q\delta(\boldsymbol{r} - \boldsymbol{v}t) \exp[-\mathrm{i}(\boldsymbol{k}_\perp \cdot \boldsymbol{r}_\perp - \omega t)]\, \mathrm{d}\boldsymbol{r}_\perp\, \mathrm{d}t$$

$$= \frac{q \exp(\mathrm{i}\omega z/v)}{(2\pi)^{3/2} v}$$

we obtain the following equation for the vacuum potential φ_1:

$$\Delta\varphi_1 = -k^2\varphi_1 + \frac{\mathrm{d}^2\varphi_1}{\mathrm{d}z^2} = -4\pi\rho_{ex} = -\frac{2q \exp(\mathrm{i}\omega z/v)}{(2\pi)^{1/2} v}$$

so that

$$\varphi_1 = A \exp(kz) + \frac{2q \exp(\mathrm{i}\omega z/v)}{(2\pi)^{1/2}(k^2 + \omega^2/v^2)v}. \tag{5.17}$$

Here and in what follows k means the absolute value of the vector $\boldsymbol{k}_\perp$, and only the decreasing solution at $z \to -\infty$ is kept in (5.17). In the same way in the dielectric ($z > 0$)

$$\varphi_2 = B \exp(-kz) + \frac{2q \exp(\mathrm{i}\omega z/v)}{(2\pi)^{1/2} v\varepsilon(\omega)(k^2 + \omega^2/v^2)}. \tag{5.18}$$

The as yet unknown constants A and B in (5.17) and (5.18) have to be determined from the boundary conditions at $z = 0$: the continuity of the potential φ and the normal component of the electric induction $\mathscr{D} = -\varepsilon\nabla\varphi$. After simple calculations we obtain that

$$A = -\frac{2q(\varepsilon - 1)}{(2\pi)^{1/2} v(\varepsilon + 1)(k^2 + \omega^2/v^2)} \qquad B = \frac{2q(\varepsilon - 1)}{(2\pi)^{1/2} v(k^2 + \omega^2/v^2)\varepsilon(\varepsilon + 1)}.$$

The following procedure has to be carried out. Knowing the potentials (5.17) and (5.18), it is possible to obtain the electric field, and hence the drag force, acting on the charge. We then need to separate the contribution connected with the excitation of surface waves from the net energy losses.

It is worth mentioning first that we are interested only in the force resulting from the polarization of a dielectric medium. The corresponding potential (we denote this by $\tilde{\varphi}$) may be easily found from (5.17) and (5.18) by subtraction of the potential created by the charge itself. Since the latter is described by the same expressions as in (5.17) and (5.18) but with $\varepsilon(\omega) \equiv 1$

(a charge moving in a vacuum)

$$\tilde{\varphi}_1 = A\,e^{kz} \qquad \tilde{\varphi}_2 = B\,e^{-kz} + \frac{2q\,e^{i\omega z/v}}{(2\pi)^{1/2}v(k^2+\omega^2/v^2)}(\varepsilon^{-1}(\omega)-1). \quad (5.19)$$

Then the drag force is determined by the electric field component $\tilde{E}_z$, for which it follows from (5.19) that

$$\tilde{E}_z^{(1)}(\boldsymbol{k}_\perp, \omega, z) = -kA\,e^{kz}$$

$$\tilde{E}_z^{(2)}(\boldsymbol{k}_\perp, \omega, z) = kB\,e^{-kz} - \frac{i\omega}{v}\frac{2q\,e^{i\omega z/v}[\varepsilon^{-1}(\omega)-1]}{(2\pi)^{1/2}v(k^2+\omega^2/v^2)}.$$

After inverse Fourier transformation it now becomes possible to obtain the electric field $\tilde{E}_z$ $(\boldsymbol{r}_\perp = 0,\, t = z/v)$ acting on the moving charge:

$$\tilde{E}_z^{(1)}\,(\boldsymbol{r}_\perp = 0, t = z/v) = -(2\pi)^{-1/2}\int_0^\infty k^2\,dk \int_{-\infty}^{+\infty} A\,e^{(k-i\omega/v)z}\,d\omega$$

$$\tilde{E}_z^{(2)}(\boldsymbol{r}_\perp = 0, t = z/v) = (2\pi)^{-1/2}\int_0^\infty k\,dk \int_{-\infty}^{+\infty} d\omega \quad (5.20)$$

$$\times\left(kB\,e^{-(k+i\omega/v)z} - \frac{i\omega}{v}\frac{2q(\varepsilon^{-1}-1)}{(2\pi)^{1/2}v(k^2+\omega^2/v^2)}\right).$$

So the net energy losses W are equal to

$$W = -\int_{-\infty}^{+\infty} dz\, q\tilde{E}_z(\boldsymbol{r}_\perp = 0, t = z/v).$$

The second term in equation (5.20) for $\tilde{E}_z^{(2)}$ does not depend on z and so describes the energy losses in a homogeneous medium. Hence the 'transit' part of the energy lost by the moving charge is connected with the two other terms in (5.20) that are tending to zero for $z \to \pm\infty$: thus we obtain, after integration over z, that

$$W_{tr} = -\frac{q^2}{\pi v}\int_0^\infty k^2\,dk \int_{-\infty}^{+\infty} d\omega \frac{[\varepsilon(\omega)-1]\{(k-i\omega/v)^{-1} + [\varepsilon(\omega)(k+i\omega/v)]^{-1}\}}{[\varepsilon(\omega)+1](k^2+\omega^2/v^2)}. \quad (5.21)$$

What is the contribution of the surface waves here? It follows from the problem 15 that in the non-relativistic limit the frequency of the surface wave is determined by the condition that $\varepsilon(\omega_0) = -1$. So this contribution is connected with the poles of the integrand (5.21) at the frequencies $\omega = \pm\omega_0$, where $\varepsilon(\omega)+1=0$. Therefore, in what follows, we may put $\varepsilon=-1$ everywhere except the poles. Then we obtain from (5.21) that

$$W_{\mathrm{SW}}=\frac{4iq^2}{\pi v^2}\int_0^{\infty}k^2\,\mathrm{d}k\int_{-\infty}^{+\infty}\frac{\omega\,\mathrm{d}\omega}{[\varepsilon(\omega)+1](k^2+\omega^2/v^2)} \tag{5.22}$$

It is seen now that W_{SW}—the energy of the excited surface waves, is determined by the imaginary part of the integral in (5.22). In a transparent medium the latter may be calculated in the way used in problem 32 for Cherenkov radiation. Let us introduce an infinitesimal imaginary part for the dielectric permeability $\varepsilon(\omega)$. Then the result is reduced to the calculation of two half-residues around the poles in (5.22) (figure 5.3). Indeed, in the

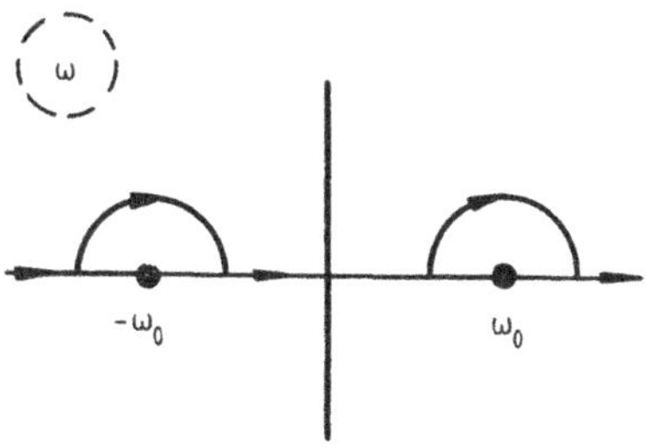

Figure 5.3. The 'passing rule' for the integration in (5.22).

vicinity of ω_0 the secular denominator $\varepsilon(\omega)+1$ may be approximated as

$$\varepsilon(\omega)+1\approx\varepsilon(\omega_0)+\left(\frac{\partial\varepsilon}{\partial\omega}\right)_{\omega_0}(\omega-\omega_0)+\mathrm{i}\varepsilon''(\omega_0)+1$$
$$=\left(\frac{\partial\varepsilon}{\partial\omega}\right)_{\omega_0}(\omega-\omega_0)+\mathrm{i}\varepsilon''.$$

So the account of ε'' results in the shifting of a pole from the real axis to the point $\omega=\omega_0-\mathrm{i}\gamma$, where $\gamma=\varepsilon''(\omega_0)/(\partial\varepsilon/\partial\omega)_{\omega_0}$. For a thermodynamic equilibrium state of a medium the quantities $\varepsilon''(\omega_0)$ and $(\partial\varepsilon/\partial\omega)_{\omega_0}$ are positive at $\omega_0>0$ (the latter of these determines the sign of the energy of the surface wave, see problem 19). Hence $\gamma>0$, the pole of $\omega=\omega_0$ is shifted down, so it needs be passed from above (figure 5.3). Since the real $(\varepsilon(\omega))$ and imaginary $(\varepsilon''(\omega))$ parts of dielectric permeability are, respectively, the even and odd functions of ω on the real axis, the same procedure at the point $\omega=-\omega_0$ gives the same result: shifting the pole down. Therefore, only these two half-residues contribute to the imaginary part of the integrand

(5.22), and

$$W_{\mathrm{SW}} = \frac{8q^2\omega_0}{v^2(\partial\varepsilon/\partial\omega)_{\omega_0}} \int_0^\infty \frac{k^2 \mathrm{d}k}{(k^2 + \omega_0^2/v^2)^2}.$$

This integrand expresses the spectral distribution of the energy for excited surface waves, so the characteristic wave length for these oscillations is $\lambda \sim v/\omega_0$, and their net energy

$$W_{\mathrm{SW}} = 2\pi q^2/v(\partial\varepsilon/\partial\omega)_{\omega_0}.$$

Chapter 6

Non-linear Interaction of Waves

Recommended textbooks:
Bloembergen N *Nonlinear Optics* (New York: Benjamin)
Shen Y R *The Principles of Nonlinear Optics* (New York: Wiley)
Tsytovich V N *Nonlinear Effects in Plasma* (New York: Plenum)

Electromagnetic equations in a continuous medium are, generally speaking, non-linear. Therefore, the expression (2.1) for the electric current in terms of the linear conductivity operator is in fact the first term in the expansion in a power-type series of the electromagnetic field. In this approximation (linear electrodynamics) arbitrary electrodynamic perturbation can be expressed as a superposition of eigenmodes (electromagnetic waves), described by equations (2.9) and (2.10), with each wave evolving independently of the others. If a non-linearity (the terms in the series after the first) are taken into account, a weak interaction between the modes is provided, so an exchange of energy between different waves now takes place (while the total energy must be conserved if the medium is non-dissipative). In the case of a weak non-linearity the main non-linear process is a *resonant three-wave interaction*. This process is possible if the wave vectors and the frequencies of the three waves (1, 2, 3) involved in the interaction satisfy the following conditions:

$$\boldsymbol{k}_1 = \boldsymbol{k}_2 + \boldsymbol{k}_3 \qquad \omega(\boldsymbol{k}_1) = \omega(\boldsymbol{k}_2) + \omega(\boldsymbol{k}_3) \tag{6.1}$$

where the frequencies $\omega(\boldsymbol{k})$ are determined by the linear dispersion equation (2.10) and are considered here and in what follows as positive. The equations for the amplitudes of these waves take the form:

$$\dot{c}_1 = V_1 c_2 c_3 \qquad \dot{c}_2 = V_2 c_1 c_3^* \qquad \dot{c}_3 = V_3 c_1 c_2^*. \tag{6.2}$$

It is convenient to normalize the amplitudes $c(\boldsymbol{k})$ in such a way that the energy of a wave $W(\boldsymbol{k})$ described by the formula (2.14) becomes equal to

$$W(\boldsymbol{k}) = \pm |c(\boldsymbol{k})|^2 \hbar\omega(\boldsymbol{k}) \tag{6.3}$$

where the $\pm$ signs correspond to a wave of positive or negative energy, and $\hbar$ is the Planck constant. Although all the processes under consideration here are classical, the introduction of the quantum constant $\hbar$ helps to give a very simple and obvious interpretation of the non-linear interaction of the waves. Indeed, if $\hbar\omega$ and $\hbar\boldsymbol{k}$ are the energy and the momentum of a single quantum, the conditions (6.1) reduce to the energy and momentum conservation at the elementary level of the three-wave interaction: the decay of a quantum 1 into quanta 2 and 3 (and the reverse process of the merging of quanta 2 and 3 into 1). It now follows from the definition (6.3) that $|c(\boldsymbol{k})|^2 \equiv N(\boldsymbol{k})$ becomes the number of quanta (*quasiparticles*) with the momentum $\hbar\boldsymbol{k}$ (and the energy $\hbar\omega$). The quantities V in the equations (6.2) are called the *matrix elements* of the three-wave interaction. If all the three interacting waves have positive energy, the following symmetry relations for V hold:

$$V_2 = V_3 = -V_1^*. \tag{6.4}$$

It is easy to check with the help of (6.2) and (6.3) that the conservation of the total energy (and momentum) is now incorporated,

$$W = \sum_{\boldsymbol{k}} W(\boldsymbol{k}) = \sum_{\boldsymbol{k}} N(\boldsymbol{k})\hbar\omega(\boldsymbol{k}) = \text{const.}$$

If one of the interacting waves (say, wave 1) has a negative energy we have, instead of (6.4),

$$V_2 = V_3 = V_1^*. \tag{6.5}$$

The total wave energy is still conserved, but it now takes the form:

$$W = \sum_{\boldsymbol{k}} N(\boldsymbol{k})\hbar\omega(\boldsymbol{k}) - \sum_{\boldsymbol{k}}{}' N(\boldsymbol{k})\hbar\omega(\boldsymbol{k})$$

where the waves with negative energy are summed with the sign minus.

In addition to the conservation of the total energy and momentum some other conservation laws connected with the occupation numbers $N(\boldsymbol{k})$ of the quasiparticles are valid for the three-wave interaction; as follows from (6.2) and (6.4)

$$N_1 + N_2 = \text{const} \qquad N_1 + N_3 = \text{const} \qquad N_2 - N_3 = \text{const.} \tag{6.6}$$

These are the so-called *Manley–Rowe relations*. Their physical sense becomes obvious from a quantum point of view. At the elementary level of the three-wave interaction the quasiparticle of type 1 decays into quasiparticles of types 2 and 3, so the decrease in N_1 is accompanied by an increase in N_2 and N_3 of the same magnitude (the same argument is also valid for the reverse process—the merging of quanta 2 and 3 into the quantum 1).

If one of the interacting waves (say, wave 1) has a negative energy, so the symmetry conditions (6.5) hold, and the Manley–Rowe relations become:

$$N_1 - N_2 = \text{const} \qquad N_1 - N_3 = \text{const} \qquad N_2 - N_3 = \text{const.} \tag{6.7}$$

Thus the elementary interaction is now a simultaneous creation (or annihilation) of three quasiparticles.

Problem 37

Determine whether a resonant three-wave interaction for waves with a given dispersion law $\omega(\boldsymbol{k})$ in an isotropic medium is possible.

The three-wave interaction is possible if the resonant conditions (6.1) may be satisfied. For an isotropic medium a wave frequency is independent of the direction of the wave vector, so $\omega(\boldsymbol{k}) \equiv \omega(k)$ and the corresponding conditions may be found simply by graphical means (see figure 6.1 (*a*)). It follows from (6.1) that the wave vectors for interacting waves must be coplanar. Let their common plane be the (k_x, k_y) plane. Then the function

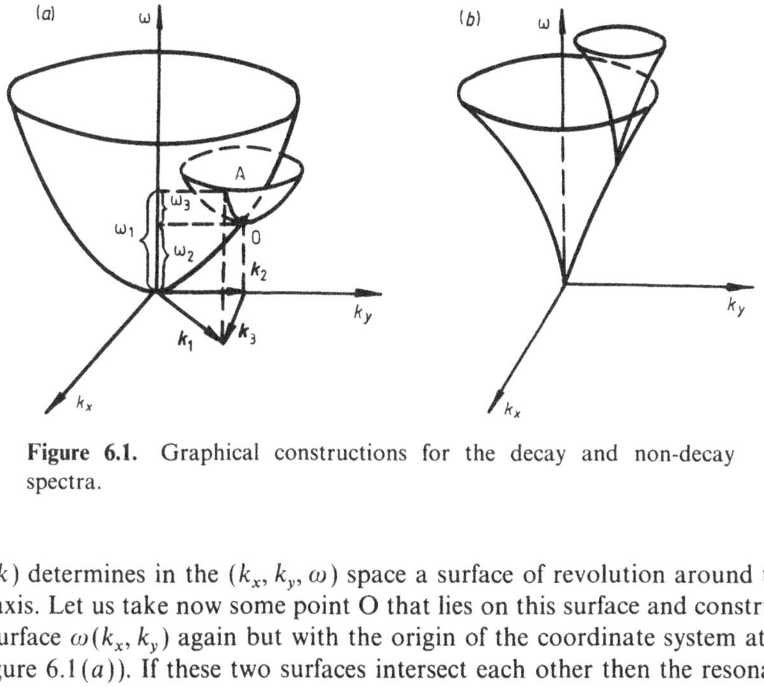

Figure 6.1. Graphical constructions for the decay and non-decay spectra.

$\omega(k)$ determines in the (k_x, k_y, ω) space a surface of revolution around the ω-axis. Let us take now some point O that lies on this surface and construct a surface $\omega(k_x, k_y)$ again but with the origin of the coordinate system at O (figure 6.1 (*a*)). If these two surfaces intersect each other then the resonant conditions (6.1) may be satisfied, and such a spectrum $\omega(k)$ is called a *decay* spectrum. Indeed, consider any point A lying on the intersection line and the projections of O and A on the (k_x, k_y) plane and the ω-axis. As is seen from the figure, three vectors are generated in this way, $\boldsymbol{k}_1$, $\boldsymbol{k}_2$ and $\boldsymbol{k}_3$, and the corresponding frequencies ω_1, ω_2 and ω_3 satisfy the resonant conditions (6.1). An example of a *non-decay* spectrum is shown in figure 6.1 (*b*). In this case the surfaces constructed above do not intersect, so the conditions (6.1) cannot be satisfied.

Problem 38

Determine the possibility of a resonant three-wave interaction for the following types of waves:

(a) $\omega = (\alpha k^3/\rho)^{1/2}$—capillary waves on a surface of a deep fluid (ρ is the density of the fluid, and α is the capillary coefficient);

(b) $\omega = (gk)^{1/2}$—gravitational waves on the surface of a deep fluid;

(c) $\omega = (\omega_{pe}^2 + k^2c^2)^{1/2}$—electromagnetic waves in a plasma;

(d) $\omega = kc_s/(1 + k^2a^2)^{1/2}$—ion sound waves in a plasma (c_s is the sound velocity and a is the Debye radius);

(e) collective excitations in liquid helium with the spectrum shown in figure 6.2 (Landau spectrum). For $k \ll k_0$, $\omega \approx ku$ (phonons), and $\omega \approx \omega_0 + (k - k_0)^2/2\mu$ for $|k - k_0| \ll k_0$ (rotons).

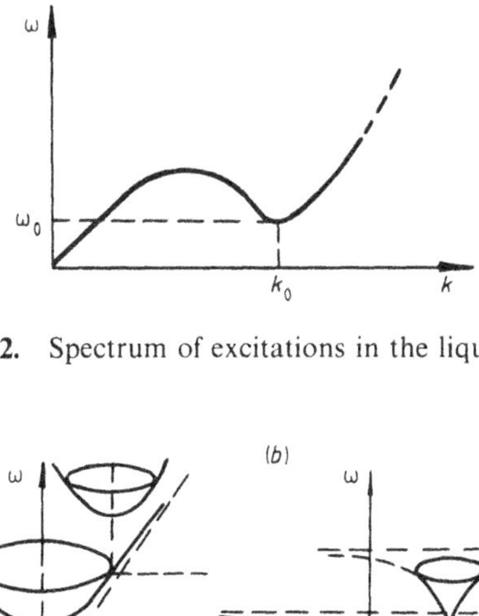

Figure 6.2. Spectrum of excitations in the liquid helium.

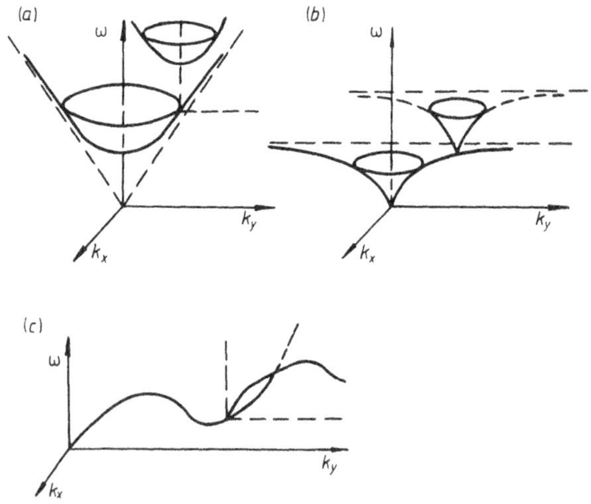

Figure 6.3. Examples of decay and non-decay spectra.

The method described in the previous problem may be applied here. It is easy to see that spectra (a) and (b) are exactly the same as shown in figure 6.1. So, capillary waves have a decay spectrum, and gravitational waves have a non-decay one. By constructing the figures for other types of waves (figure 6.3) we conclude that electromagnetic and sound waves in a plasma have non-decay spectra. In liquid helium a decay of a roton into a phonon and another roton is possible. For this spectrum the intersection takes place in the plane of the figure, so one-dimensional decay processes are possible, when all three interacting waves have collinear wave vectors (unlike, for example, capillary waves, for which one-dimensional decays are prohibited).

Problem 39

Find the minimum frequency of a wave that can decay into two other waves, if the dispersion law is $\omega(k) = \omega_0 + \alpha k^2$.

The answer is not $2\omega_0$, because not only energy conservation but also momentum conservation must be satisfied. It follows from the conditions (6.1) and the dispersion equation $\omega(k) = \omega_0 + \alpha k^2$, that $(\boldsymbol{k}_2 \cdot \boldsymbol{k}_3) = \omega_0/2\alpha$. So the problem is reduced to finding a minimum for the sum $(k_2^2 + k_3^2)$ under a given scalar product of these vectors $(\boldsymbol{k}_2 \cdot \boldsymbol{k}_3)$. It is clear that the minimum will be achieved when $\boldsymbol{k}_2$ and $\boldsymbol{k}_3$ are parallel. Then $k_3 = \omega_0/2\alpha k_2$, and $k_2^2 + k_3^2 = k_2^2 + \omega_0^2/4\alpha^2 k_2^2$. So it has a minimum at $k_2 = k_3 = (\omega_0/2\alpha)^{1/2}$. Thus the minimum frequency of a decaying wave is $\omega_{\min} = 3\omega_0$.

Problem 40

A wave l with the spectrum $\omega_l = \omega_0 + \alpha k^2$ decays according to the scheme $l \to l' + s$ into a wave l′ of the same type and a wave s with the spectrum $\omega_s = kc_s$. Find the minimum possible frequency of the decaying wave.

By plotting the corresponding graphs (as in problem 37) we see that two different cases for this decay are possible (figure 6.4). In the case (a) when the group velocity of the wave l′ is less than the sound velocity c_s, the wave l has a group velocity larger than c_s. If the frequency of the wave l′ is so large that its group velocity exceeds c_s (case b), it is possible for a wave l to have a frequency very close to that of the wave l′. Since for a wave of type l the group velocity is a monotonically increasing function of frequency, we may conclude that ω_l would be a minimum when the group velocity of the wave l′ is equal to c_s. Then $\omega_l \to \omega_{l'}$ and $\omega_s \to 0$, so that $(\partial\omega_l/\partial k)_{k_0} = 2\alpha k_0 = c_s$, $k_0 = c_s/2\alpha$ and $\omega_{\min} = \omega_0 + \alpha k_0^2 = \omega_0 + c_s^2/4\alpha$.

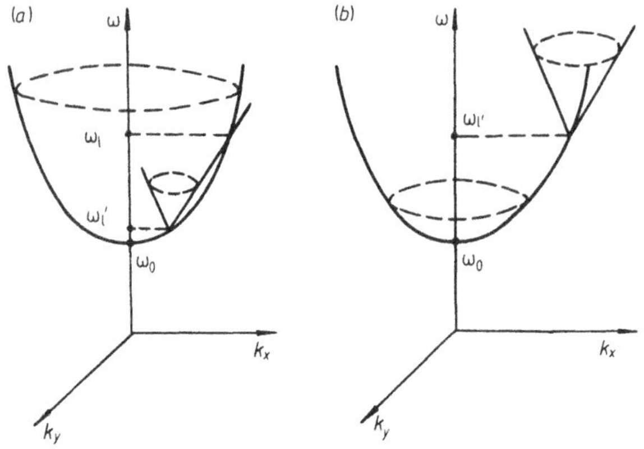

Figure 6.4. Decay of an l wave into an l′ wave and an s wave.

Problem 41

Find the growth rate of the decay instability arising for the three-wave interaction of waves with positive energy.

The phenomenon of the decay instability consists in a stationary state of the system (6.2) being unstable when only a wave with the maximum frequency (wave 1 in our case) is excited while the amplitudes of the other two waves, c_2 and c_3, are equal to zero. Under small perturbations the energy of wave 1 is transferred to waves 2 and 3.

Thus at the initial moment $c_1 = c_0$ and $c_2 = c_3 = 0$. Then introduce the small perturbations $c_1 = c_0 + a_1\, e^{\gamma t}$, $c_2 = a_2\, e^{\gamma t}$ and $c_3 = a_3\, e^{\gamma t}$ which depend exponentially on time. By substitution of these expressions for the amplitudes in equations (6.2) and taking into account (6.4) we obtain in the linear approximation:

$$a_1 = 0 \qquad \gamma a_2 = -V^* c_0 a_3 \qquad \gamma a_3 = -V^* c_0 a_2^*.$$

It follows now that $|\gamma|^2 = |V|^2 |c_0|^2$, so the exponentially growing solution with Re $\gamma > 0$ always exists. It is worth noting here that the wave 1 is specific because it has a maximum frequency. It is easy to check directly from equations (6.2) that the stationary states when only wave 2 or wave 3 is excited, are stable. In this case the small perturbations have an oscillatory behaviour.

At the decay instability of wave 1 the exponential growth of perturbations take place only at the initial stage of the process, until $a \ll c_0$. The amplitudes of waves 2 and 3 grow until the energy of these waves becomes of the same order as the energy of wave 1. Then the reverse transfer of energy occurs, and so on. Such a behaviour of the amplitudes may be predicted at once without solving equations (6.2), but based on the conservation of the total energy of the waves and the Manley–Rowe relations. For the waves with positive energy it gives an upper limit for the occupation numbers $N(\mathbf{k})$ of each wave.

The situation becomes rather different if a wave with negative energy is involved in the interaction. In this case the unbounded growth of all the waves is possible without any contradiction with the conservation laws (see the next problem).

Problem 42

Show that if a wave with negative energy takes part in a three-wave interaction, the so-called 'explosive' instability is possible, when the amplitudes of the waves tend to infinity at finite time. Consider the behaviour of amplitudes near the moment of the 'explosion'.

Let us consider the simplest case when wave 1 with a maximum frequency has a negative energy, and at the initial moment ($t = 0$) only waves 2 and 3 are excited: $c_2(0) = c_3(0) = c_0$; $c_1(0) = 0$. We shall also assume that the initial amplitude c_0 and the matrix element V are real. Then, as follows from (6.2) and (6.5), at any moment all the amplitudes are real. From the Manley–Rowe relations (6.7) we now have:

$$c_2(t) = c_3(t) \equiv c(t) \qquad c_1^2(t) - c_2^2(t) = \text{const} = -c_0^2$$
$$c^2(t) = c_1^2(t) + c_0^2.$$

After that the first equation (6.2) takes the form:

$$\mathrm{d}c_1/\mathrm{d}t = Vc_2c_3 = Vc^2 = V(c_1^2 + c_0^2)$$

and may be integrated simply: $c_1(t) = c_0 \tan(Vc_0t)$. As a result the amplitudes of all three waves become infinite at the moment $t = t_0 = \pi/2Vc_0$. In the vicinity of this moment their behaviour is: $c_{1,2,3} \propto (t_0 - t)^{-1}$.

Problem 43

Obtain non-linear equations describing the interaction of electromagnetic and Langmuir waves in a cold-electron plasma (see problem 7).

The permittivity of such a plasma is $\varepsilon(\omega) = 1 - \omega_{\mathrm{pe}}^2/\omega^2$, where $\omega_{\mathrm{pe}}^2 = 4\pi n_0 e^2/m$ (n_0, e, m are the density, the charge and the mass of the electrons, respectively), so the dispersion equations are $\omega(k) = (\omega_{\mathrm{pe}}^2 + k^2c^2)^{1/2}$ for an electromagnetic wave and $\omega(k) = \omega_{pe}$ for a Langmuir wave. Although each of these spectra is of a non-decay type (see problem 38), a three-wave interaction is possible here if, for instance, waves 1 and 2 are electromagnetic, and wave 3 is a Langmuir wave. Let us consider the simplest case when all the waves propagate in the same direction (say, along the z-axis), and transverse electromagnetic waves have a linear polarization with the electric field along the x-axis (so that only E_x and B_y are present). The resonant conditions: $k_1 = k_2 + k_3$; $\omega(k_1) = \omega(k_2) + \omega(k_3)$ may now be easily satisfied if the frequency of a decaying electromagnetic wave $\omega_1 \equiv \omega(k_1)$ exceeds $2\omega_{pe}$ (k_1 and k_2 are the wave vectors of the electromagnetic waves, and k_3 is the wave vector of the Langmuir wave). The Maxwell equations and the electron's equation of motion are:

$$\operatorname{curl} \boldsymbol{E} = -\frac{1}{c}\frac{\partial \boldsymbol{B}}{\partial t}$$

$$\operatorname{curl} \boldsymbol{B} = \frac{1}{c}\frac{\partial \boldsymbol{E}}{\partial t} + \frac{4\pi e}{c}(n_0 + n)\boldsymbol{v} = \frac{1}{c}\frac{\partial \boldsymbol{E}}{\partial t} + \frac{4\pi n_0 e}{c}\boldsymbol{v} + \frac{\boldsymbol{v}}{c}\operatorname{div} \boldsymbol{E} \tag{6.8}$$

$$\frac{\partial \boldsymbol{v}}{\partial t} + (\boldsymbol{v}\cdot\nabla)\boldsymbol{v} = \frac{e}{m}\left(\boldsymbol{E} + \frac{1}{c}(\boldsymbol{v}\times\boldsymbol{B})\right)$$

where n is the deviation of the density of electrons from their equilibrium value n_0 (we also use here the Poisson equation, so $n = \operatorname{div} \boldsymbol{E}/4\pi e$). In our case only the following quantities differ from zero: E_x, B_y, v_x, E_z and v_z. By representing each of these as a Fourier series, say,

$$E_x(z, t) = \sum_k E_{xk}(t)\,\mathrm{e}^{ikz}$$

we obtain from (6.8) the following equations:

$$\frac{\partial E_{xk}}{\partial t} = -\mathrm{i}ckB_{yk} - 4\pi n_0 e v_{xk} - \mathrm{i}\sum_{k'} k' E_{zk'} v_{x(k-k')}$$

$$\partial B_{yk}/\partial t = -\mathrm{i}ckE_{xk}$$

$$\frac{\partial v_{xk}}{\partial t} = \frac{e}{m}E_{xk} - \mathrm{i}\sum_{k'} k' v_{xk'} v_{z(k-k')} - \frac{e}{mc}\sum_{k'} B_{yk'} v_{z(k-k')} \tag{6.9}$$

$$\frac{\partial E_{zk}}{\partial t} = -4\pi n_0 e v_{zk} - \mathrm{i}\sum_{k'} k' E_{zk'} v_{z(k-k')}$$

$$\frac{\partial v_{zk}}{\partial t} = \frac{e}{m}E_{zk} - \mathrm{i}\sum_{k'} k' v_{zk'} v_{z(k-k')} + \frac{e}{mc}\sum_{k'} B_{yk'} v_{x(k-k')}.$$

In these equations the non-linearity is determined by summing over the

k' terms in (6.9). In the linear approximation these terms may be omitted, and the Langmuir wave (E_z and v_z correspond to it) and the electromagnetic wave, (E_x, B_y, v_x), evolve independently.

Problem 44

Derive the equation for the amplitude of a wave involved in the resonant three-wave interaction starting from the non-linear equations of the type (6.9).

Equations (6.9) for the Fourier components of the perturbations in a medium may be written symbolically in the form:

$$\frac{\partial A_{\alpha k}}{\partial t} - L_{\alpha\beta}(\boldsymbol{k})A_{\beta k} = \sum_{k'} T_{\alpha\beta\gamma}(\boldsymbol{k}, \boldsymbol{k}')\, A_{\beta k'} A_{\gamma(k-k')} \tag{6.10}$$

where $\boldsymbol{A}_k$ is a 'vector of state' of a medium (for instance, in the problem considered above $\boldsymbol{A} = (E_x, B_y, v_x, E_z, v_z)$. Such a representation allows us to consider the non-linear interaction of waves of any nature (not only electromagnetic waves). The matrix $L_{\alpha\beta}(\boldsymbol{k})$ in (6.10) describes the waves in a medium in the linear approximation, and the matrix $T_{\alpha\beta\gamma}(\boldsymbol{k}, \boldsymbol{k}')$ describes the non-linearity of the system. The problem is to obtain the variation in time of the wave amplitude resulting from the non-linear interaction by considering the last term as a weak perturbation.

Let us introduce the eigenvectors of a linear problem:

$$\varphi_{\alpha k}(t) = c_k \exp(-\mathrm{i}\omega_k t)\psi_{\alpha k} \tag{6.11}$$

where $-\mathrm{i}\omega_k$ is the eigenvalue of the matrix $L_{\alpha\beta}(\boldsymbol{k})$. In the absence of dissipation the frequencies ω_k must be real, so the matrix $L_{\alpha\beta}$ has pure imaginary eigenvalues and is, therefore, an antiHermitian matrix: $L_{\beta\alpha} = -L^*_{\alpha\beta}$. It is convenient to use a normalization for the vectors $\boldsymbol{\varphi}_k$ and $\boldsymbol{\psi}_k$ such that the scalar products are $\psi_{\alpha k}\psi^*_{\alpha k} = \hbar\omega_k$, and $\varphi_{\alpha k}\varphi^*_{\alpha k}$ is equal to the energy density of the wave, W_k. Thus $W_k = |c_k|^2\hbar\omega_k$, so that $|c_k|^2 \equiv n_k$ takes account of the number of quasiparticles with given moment $\hbar\boldsymbol{k}$ and energy $\hbar\omega_k$ per unit volume. Since the components of the vector $\boldsymbol{A}_k$ are physical quantities that must be real, the Fourier amplitudes c_k and c_{-k} must satisfy the condition $c_{-k} = c^*_k$ (as well as $\omega(-\boldsymbol{k}) = -\omega(\boldsymbol{k})$). So the harmonics with wave vectors and frequencies $(\boldsymbol{k}, \omega_k)$ and $(-\boldsymbol{k}, -\omega_k)$ describe the same physical wave. In what follows we consider the frequency of the wave to be a positive quantity. If the energy of the wave is negative the normalization for its amplitude will be

$$W_k = -|c_k|^2\hbar\omega_k = -n_k\hbar\omega_k.$$

The non-linearity results in the change of the amplitudes c_k as well as in a small 'deviation' of the 'unit' vector $\boldsymbol{\psi}_k$:

$$\boldsymbol{\varphi}_k = c_k(t)\exp(-\mathrm{i}\omega_k t)[\boldsymbol{\psi}_k + \delta\psi_k(t)] \qquad (6.12)$$

and it follows from the conservation of normalization that the vectors $\boldsymbol{\psi}_k$ and $\delta\psi_k$ are orthogonal: $\psi_{\alpha k}\delta\psi^*_{\alpha k} = 0$. By putting expression (6.12) for $\boldsymbol{\varphi}_k$ in equation (6.10) and taking into account that for a weak non-linearity the non-perturbed expression (6.11) may be used on the right-hand side, we have:

$$\begin{aligned}
&\dot{c}_k \exp(-\mathrm{i}\omega_k t)\psi_{\alpha k} - \mathrm{i}\omega_k c_k \exp(-\mathrm{i}\omega_k t)(\psi_{\alpha k} + \delta\psi_{\alpha k}) \\
&\quad - c_k \exp(-\mathrm{i}\omega_k t)L_{\alpha\beta}(\boldsymbol{k})(\psi_{\beta k} + \delta\psi_{\beta k}) \\
&= \sum_{k'} T_{\alpha\beta\gamma}(\boldsymbol{k}, \boldsymbol{k}')c_{k'}c_{(k-k')}\psi_{\beta k'}\psi_{\gamma(k-k')}\exp[-\mathrm{i}(\omega_{k'} + \omega_{(k-k')})t]
\end{aligned} \qquad (6.13)$$

(some terms of the higher order as we need, like $\dot{c}_k\delta\psi_k$ or $\delta\dot{\boldsymbol{\psi}}c_k$, are omitted in (6.13)). Since $\boldsymbol{\psi}_k$ is the eigenvector of the matrix $L_{\alpha\beta}(\boldsymbol{k})$, it follows from (6.13) that

$$\begin{aligned}
&\dot{c}_k\psi_{\alpha k} - \mathrm{i}\omega_k c_k\delta\psi_{\alpha k} - c_k L_{\alpha\beta}\delta\psi_{\beta k} \\
&\qquad = \sum_{k'} T_{\alpha\beta\gamma}(\boldsymbol{k}, \boldsymbol{k}')\psi_{\beta k'}\psi_{\gamma(k-k')}c_{k'}c_{(k-k')}\exp[\mathrm{i}(\omega_k - \omega_{k'} - \omega_{(k-k')})t].
\end{aligned} \qquad (6.14)$$

From this vector equation and the orthogonality condition $\psi_k\delta\boldsymbol{\psi}^*_k = 0$ the amplitude evolution $\dot{c}_k$ and the correction $\delta\psi_k$ to the 'unit' vector may be obtained. However, since we are interested only in $\dot{c}_k$, it is convenient to derive one scalar equation from (6.14). To do this, we take the scalar product of both sides of (6.14) with the vector $\psi^*_{\alpha k}$. Then the second term in the left-hand side of (6.14) vanishes because of orthogonality, and the third term may be transformed as follows:

$$L_{\alpha\beta}\delta\psi_\beta\psi^*_\alpha = -L^*_{\beta\alpha}\psi^*_\alpha\delta\psi_\beta = -\mathrm{i}\omega\psi^*_\beta\delta\psi_\beta = 0$$

(we use here that $L_{\alpha\beta}$ is an antiHermitian matrix and ψ_α is its eigenvector). Thus the correction $\delta\psi_k$ cancels. The main contribution to the right-hand side of (6.14) arises from the resonant terms, when $\omega_k - \omega_{k'} - \omega_{(k-k')} = 0$. Otherwise the exponent in (6.14) oscillates very rapidly and cannot provide the effective interaction of waves. Noting the sum over only resonant terms such as $\tilde{\Sigma}_{k'}$, we finally obtain:

$$\dot{c}_k = \tilde{\sum_{k'}} V(\boldsymbol{k}, \boldsymbol{k}')c_{k'}c_{(k-k')} \qquad v(\boldsymbol{k}, \boldsymbol{k}') = \frac{T_{\alpha\beta\gamma}(\boldsymbol{k}, \boldsymbol{k}')\psi^*_{\alpha k}\psi_{\beta k'}\psi_{\gamma(k-k')}}{\hbar\omega_k} \qquad (6.15)$$

Problem 45

Find the matrix elements for the resonant three-wave interaction of electromagnetic and Langmuir waves in a plasma (see problem 43).

The scheme developed in the previous problem may be applied for this particular process: one-dimensional decay of an electromagnetic wave on another electromagnetic wave and a Langmuir wave. The 'vector of state' A may now be determined as follows:

$$A = (E_x, B_y, (4\pi n_0 m)^{1/2} v_x, E_z, (4\pi n_0 m)^{1/2} v_z).$$

Then the system (6.9) takes the form:

$$\frac{\partial A_{1k}}{\partial t} = -ickA_{2k} + \omega_{\text{pe}} A_{3k} + \mathrm{i}\frac{2\sqrt{\pi}e}{m\omega_{\text{pe}}}\sum_{k'}(k-k')A_{3k'}A_{4(k-k')}$$

$$\partial A_{2k}/\partial t = -ickA_{1k}$$

$$\frac{\partial A_{3k}}{\partial t} = -\omega_{\text{pe}} A_{1k} - \frac{2\sqrt{\pi}e}{mc}\sum_{k'} A_{2k'}A_{5(k-k')} + \mathrm{i}\frac{2\sqrt{\pi}e}{m\omega_{\text{pe}}}\sum_{k'} k' A_{3k'}A_{5(k-k')} \tag{6.16}$$

$$\frac{\partial A_{4k}}{\partial t} = \omega_{\text{pe}} A_5 + \mathrm{i}\frac{2\sqrt{\pi}e}{m\omega_{\text{pe}}}\sum_{k'} k' A_{4k'}A_{5(k-k')}$$

$$\frac{\partial A_{5k}}{\partial t} = -\omega_{\text{pe}} A_4 + \frac{2\sqrt{\pi}e}{mc}\sum_{k'} A_{2k'}A_{3(k-k')} + \mathrm{i}\frac{2\sqrt{\pi}e}{m\omega_{pe}}\sum_{k'} k' A_{5k'}A_{5(k-k')}.$$

In accordance with a general rule the linear matrix $L_{\alpha\beta}$ is an antiHermitian one and has the following non-zero components:

$$L_{12} = -L_{21}^* = -ick \qquad L_{13} = -L_{31}^* = \omega_{\text{pe}} \qquad L_{45} = -L_{54}^* = \omega_{\text{pe}}.$$

Its eigenvectors correspond to transverse electromagnetic waves (t) and longitudinal Langmuir waves (l):

$$\varphi_\alpha^{(l)} = c^{(l)}\exp(-\mathrm{i}\omega_{\text{pe}} t)\psi_\alpha^{(l)} \qquad \varphi_\alpha^{(t)} = c^{(t)}\exp(-\mathrm{i}\omega^{(t)} t)\psi_\alpha^{(t)}$$

$$\omega^{(t)} = (\omega_{\text{pe}}^2 + k^2c^2)^{1/2}$$

where the properly normalized 'unit' vectors $\psi^{(l)}$ and $\psi^{(t)}$ are

$$\psi^{(l)} = \left(\frac{\hbar\omega_{\text{pe}}}{2}\right)^{1/2}(0, 0, 0, 1, -\mathrm{i})$$

$$\psi^{(t)} = \left(\frac{\hbar\omega^{(t)}}{2}\right)^{1/2}\left(1, \sqrt{\varepsilon(\omega^{(t)})}, -\frac{\mathrm{i}\omega_{\text{pe}}}{\omega^{(t)}}, 0, 0\right). \tag{6.17}$$

Let the electromagnetic wave with the wave vector k_1, the frequency $\omega_1 = \omega^{(t)}(k_1)$ and with the amplitude c_1 now decay into another electromagnetic wave (with wave vector k_2, frequency $\omega_2 = \omega^{(t)}(k_2)$ and amplitude c_2) and Langmuir wave (with wave vector $k_3 = k_1 - k_2$, frequency $\omega_3 \equiv \omega_{pe} = \omega_1 - \omega_2$ and amplitude c_3). It follows from (6.16) that for this case the matrix of the non-linear interaction $T_{\alpha\beta\gamma}$ has the following non-zero components:

$$T_{134}(k, k') = \mathrm{i}\frac{2\sqrt{\pi}e}{m\omega_{pe}}(k - k') \qquad T_{325}(k, k') = -\frac{2\sqrt{\pi}e}{mc}$$

$$T_{335}(k, k') = \mathrm{i}\frac{2\sqrt{\pi}e}{m\omega_{pe}}k' \qquad T_{445}(k, k') = \mathrm{i}\frac{2\sqrt{\pi}e}{m\omega_{pe}}k'$$

$$T_{523}(k, k') = \frac{2\sqrt{\pi}e}{mc} \qquad T_{555} = \mathrm{i}\frac{2\sqrt{\pi}e}{m\omega_{pe}}k'.$$

Thus according to the general formula (6.15) and taking into account the 'unit' vectors $\psi_\alpha^{(l,t)}$ from (6.17) we find:

$$\frac{\mathrm{d}c_1}{\mathrm{d}t} = V_1 c_2 c_3, \qquad (6.18)$$

where

$$\begin{aligned} V_1 &= \frac{1}{\hbar\omega_1}[T_{134}(k_1, k_2)\psi^*_{1k_1}\psi_{3k_2}\psi_{4k_3} + T_{325}(k_1, k_2)\psi^*_{3k_1}\psi_{2k_2}\psi_{5k_3} \\ &\quad + T_{335}(k_1, k_2)\psi^*_{3k_1}\psi_{3k_2}\psi_{5k_3}] \\ &= -\frac{2\sqrt{\pi}e}{\hbar m\omega_1}\frac{(\hbar^3\omega_1\omega_2\omega_{pe}/2)^{1/2}}{2}\left(-\frac{k_3}{\omega_2} + \frac{\omega_{pe}\varepsilon^{1/2}(\omega_2)}{c\omega_1} - \frac{k_2\omega_{pe}}{\omega_1\omega_2}\right) \\ &= \frac{e(k_1 - k_2)}{m}\left(\frac{\pi\hbar\omega_{pe}}{2\omega_1\omega_2}\right)^{1/2}. \end{aligned}$$

How do we derive the equations for the other amplitudes, c_2 and c_3? To do this, we note that the resonant conditions for these three waves may be written formally as: $k_2 = k_1 + (-k_3)$; $\omega_2 = \omega_1 + (-\omega_3)$. It was mentioned above that changing the sign of the wave vector and frequency is equivalent to the transition to complex-conjugate values of the amplitude and a 'unit' vector of the wave, so

$$\mathrm{d}c_2/\mathrm{d}t = V_2 c_1 c_3^* \qquad (6.19)$$

where

$$V_2 = \frac{1}{\hbar\omega_2}[T_{134}(k_2, k_1)\psi^*_{1k_2}\psi_{3k_1}\psi^*_{4k_3} + T_{325}(k_2, k_1)\psi^*_{3k_2}\psi_{2k_1}\psi^*_{5k_3}$$

$$+ T_{335}(k_2, k_1)\psi^*_{3k_2}\psi_{3k_1}\psi^*_{5k_3}]$$

$$= -\frac{e(k_1 - k_2)}{m}\left(\frac{\pi\hbar\omega_{pe}}{2\omega_1\omega_2}\right)^{1/2}.$$

Similarly, for c_3 we have

$$dc_3/dt = V_3 c_1 c_2^*$$

$$V_3 = \frac{1}{\hbar\omega_{pe}}[T_{523}(k_3, k_1)\psi^*_{5k_3}\psi_{2k_1}\psi^*_{3k_2} + T_{523}(k_3, -k_2)\psi^*_{5k_3}\psi^*_{2k_2}\psi_{3k_1}] \tag{6.20}$$

$$= V_2.$$

Thus equations (6.18)–(6.20) for the amplitudes may be written in the standard form (6.2) and (6.4) for the interaction of waves with positive energies:

$$\dot{c}_1 = Vc_2c_3 \qquad \dot{c}_2 = -V^*c_1c_3^* \qquad \dot{c}_3 = -V^*c_1c_2^*$$

$$V = \frac{e(k_1 - k_2)}{m}\left(\frac{\pi\hbar\omega_{pe}}{2\omega_1\omega_2}\right)^{1/2}. \tag{6.21}$$

Although the matrix element of an interaction V contains the quantum constant $\hbar$, this process, obviously, is purely classical. It is easy to see that the appearance of $\hbar$ in (6.21) is a result of our normalization of the amplitudes, when the occupation numbers $n_k = |c_k|^2 \propto \hbar^{-1}$. As was mentioned above such a normalization has the advantage of giving a simple interpretation of the three-wave interaction as a decay or merging of quasiparticles. From the quantum point of view equations (6.21) correspond to the classical limit of large occupation numbers, when the induced processes (rather than the spontaneous ones) dominate.

Problem 46

Derive the kinetic equation for an ensemble of resonantly interacting waves.

If a broad spectrum of waves is excited in a medium, the resonant conditions (6.1) may be satisfied for different sets of three waves, so that each wave interacts simultaneously with a large number of other waves. In this situation the statistical description becomes adequate, when the phases of waves may be considered as distributed chaotically (the so-called 'random-phase' approximation), and the evolution of the wave-like perturbations is described in terms of the spectral occupation numbers N_k. To obtain the corresponding kinetic equation for N_k we shall start from equations (6.14) and (6.15) for the amplitude c_k, now considering a continuous spectrum (so an integration over $\boldsymbol{k}'$ must be carried out):

$$\frac{\mathrm{d}c_k}{\mathrm{d}t} = \frac{1}{2}\int \mathrm{d}\boldsymbol{k}'\, V_{kk'}c_{k'}c_{k''} \exp[\mathrm{i}(\omega_k - \omega_{k'} - \omega_{k''})t] \tag{6.22}$$

where $\boldsymbol{k}'' = \boldsymbol{k} - \boldsymbol{k}'$ (the factor of a half arises here because the interchange of the wave vectors $\boldsymbol{k}'$ and $\boldsymbol{k}''$ results in the same process). Thus, the change in occupation number N_k is

$$\frac{\mathrm{d}N_k}{\mathrm{d}t} = c_k^* \frac{\mathrm{d}c_k}{\mathrm{d}t} + c_k \frac{\mathrm{d}c_k^*}{\mathrm{d}t} = \frac{1}{2}\int \mathrm{d}\boldsymbol{k}'\, V_{kk'}c_{k'}c_{k''}c_k^*\, \mathrm{e}^{\mathrm{i}(\Delta\omega)t} + \mathrm{cc} \tag{6.23}$$

$$\Delta\omega = \omega_k - \omega_{k'} - \omega_{k''}.$$

Now we need to perform in this equation the phase averaging in the amplitudes c_k, $c_{k'}$ and $c_{k''}$. To do this, we use a perturbation method representing an amplitude as a series in terms of the interaction operator V: $c_k = c_k^{(0)} + c_k^{(1)} + \cdots$ The chaotic phase state means that there are no phase correlations in the zeroth-order approximation, that is

$$\langle c_k^{(0)}\rangle = 0 \qquad \langle c_k^{(0)}c_{k'}^{(0)*}\rangle = N_k\delta(\boldsymbol{k} - \boldsymbol{k}')$$
$$\langle c_k^{(0)}c_{k'}^{(0)}\rangle = N_k\delta(\boldsymbol{k} + \boldsymbol{k}') \tag{6.24}$$

(the sign $\langle\ \rangle$ means averaging over phases, and the relation $c_k^* = c_{-k}$ is used here). Thus in the zeroth-order approximation the right-hand side of equation (6.23) becomes zero. So we need to know the next term in the series, $c_k^{(1)}$, to obtain $\dot{N}_k$. It follows from (6.22) that

$$\frac{\mathrm{d}c_k^{(1)}}{\mathrm{d}t} \approx \frac{1}{2}\int \mathrm{d}\boldsymbol{k}'\, V_{kk'}c_{k'}^{(0)}c_{k''}^{(0)}\, \mathrm{e}^{-\mathrm{i}(\Delta\omega)t}. \tag{6.25}$$

After integration of (6.25) over time from $t = -\infty$ to a given moment (we may put this as $t = 0$) and introducing the adiabatically slow switching on of the interaction ($V \propto \exp(-\gamma^2t^2)$, so that $c_k^{(1)} = 0$ at $t \to -\infty$, we obtain from (6.25):

$$\begin{aligned} c_k^{(1)} &= \frac{1}{2}\int \mathrm{d}\boldsymbol{k}'\, V_{kk'}c_{k'}^{(0)}c_{k''}^{(0)} \int_{-\infty}^{0} \exp[-\mathrm{i}(\Delta\omega)t - \gamma^2t^2]\,\mathrm{d}t \\ &= \int \mathrm{d}\boldsymbol{k}'\, V_{kk'}c_{k'}^{(0)}c_{k''}^{(0)} \frac{\sqrt{\pi}}{2\gamma}\exp[-(\Delta\omega)^2/4\gamma^2] \end{aligned} \tag{6.26}$$

(the amplitudes $c_{k'}^{(0)}$ and $c_{k''}^{(0)}$ must be considered here as independent of time, since taking into account their variation leads to higher-order corrections in the interaction). Since

$$\lim_{\gamma\to 0}\frac{\exp[-(\Delta\omega)^2/4\gamma^2}{2\gamma\sqrt{\pi}}=\delta(\Delta\omega)$$

it follows from (6.26) that

$$c_k^{(1)}\approx\frac{\pi}{2}\int \mathrm{d}\boldsymbol{q}\; V_{kq}c_q^{(0)}c_{k-q}^{(0)}\delta(\Delta\omega). \qquad (6.27)$$

In the same way the corrections $c_{k'}^{(1)}$ and $c_{k''}^{(1)}$ may be found. By using equations (6.2) and the symmetry relations for the matrix elements (6.4) we find

$$\begin{aligned} c_{k'}^{(1)}&=-\frac{\pi}{2}\int \mathrm{d}\boldsymbol{q}\; V^*c_q^{(0)}c_{q-k'}^{(0)*}\delta(\Delta\omega)\\ c_{k''}^{(1)}&=-\frac{\pi}{2}\int \mathrm{d}\boldsymbol{q}\; V^*c_q^{(0)}c_{q.-k''}^{(0)*}\delta(\Delta\omega) \end{aligned} \qquad (6.28)$$

(we assume here that interacting waves have a positive energy, so that relations (6.4) are valid). By substitution of (6.27) and (6.28) in (6.23) and phase averaging we obtain:

$$\begin{aligned}\frac{\mathrm{d}N_k}{\mathrm{d}t}&=\frac{1}{2}\int \mathrm{d}\boldsymbol{k}'\; V\,\mathrm{e}^{\mathrm{i}(\Delta\omega)t}\langle c_{k'}^{(1)}c_{k''}^{(0)}c_k^{(0)*}+c_{k'}^{(0)}c_{k''}^{(1)}c_k^{(0)*}+c_{k'}^{(0)}c_{k''}^{(0)}c_k^{(1)*}\rangle+\mathrm{cc}\\ &=\frac{\pi}{2}\int \mathrm{d}\boldsymbol{k}'\; V\Big(\int \mathrm{d}\boldsymbol{q}\; V^*\langle c_{k'}^{(0)}c_{k''}^{(0)}c_q^{(0)*}c_{k-q}^{(0)*}\rangle-\int \mathrm{d}\boldsymbol{q}\; V^*\langle c_{k''}^{(0)}c_k^{(0)*}c_q^{(0)}c_{q-k'}^{(0)}\rangle\\ &\quad-\int \mathrm{d}\boldsymbol{q}\; V^*\langle c_{k'}^{(0)}c_k^{(0)*}c_q^{(0)}c_{q-k''}^{(0)*}\rangle\Big)\delta(\Delta\omega). \end{aligned} \qquad (6.29)$$

Since there are no phase correlations in the amplitudes of the zeroth-order approximation, the phase averaging of a product of four of them in (6.29) reduces to a product of two occupation numbers. Indeed, in the first term on the right-hand side of (6.29) a non-zero result is possible only in two cases: $\boldsymbol{q}=\boldsymbol{k}'$ or $\boldsymbol{q}=\boldsymbol{k}''$. By proceeding with the same 'pairing' in the other terms we find the kinetic equation:

$$\frac{\mathrm{d}N_k}{\mathrm{d}t}=\pi\int \mathrm{d}\boldsymbol{k}'\;|V_{kk'}|^2(N_{k'}N_{k-k'}-N_kN_{k'}-N_kN_{k-k'})\delta(\omega_k-\omega_{k'}-\omega_{k-k'}). \qquad (6.30)$$

The equations for $N_{k'}$ and $N_{k-k'}$ result at once from the Manley–Rowe relations:

$$\mathrm{d}N_{k'}/\mathrm{d}t=\mathrm{d}N_{k-k'}/\mathrm{d}t=-\mathrm{d}N_k/\mathrm{d}t. \qquad (6.31)$$

The kinetic equation (6.30) may also be derived from the quantum-mechanical correspondence principle by considering a balance of quasiparticles at an elementary three-wave interaction. Let $w(\boldsymbol{k}, \boldsymbol{k}')\,\mathrm{d}\boldsymbol{k}'$ be the probability of decay of a quasiparticle with momentum $\hbar\boldsymbol{k}$ into two other quasiparticles with momenta lying in the interval $[\hbar\boldsymbol{k}', \hbar(\boldsymbol{k}' + \mathrm{d}\boldsymbol{k}')]$ and $[\hbar(\boldsymbol{k} - \boldsymbol{k}'), \hbar(\boldsymbol{k} - \boldsymbol{k}' - \mathrm{d}\boldsymbol{k}')]$. In accordance with the detailed balancing principle[†] the reverse process (the merging of two quasiparticles into one) has the same probability. So a balance equation for the occupation number $N_{\boldsymbol{k}}$ takes the form:

$$\frac{\mathrm{d}N_{\boldsymbol{k}}}{\mathrm{d}t} = \frac{1}{2}\int w(\boldsymbol{k}, \boldsymbol{k}')\,\mathrm{d}\boldsymbol{k}'[(N_{\boldsymbol{k}} + 1)N_{\boldsymbol{k}'}N_{\boldsymbol{k}-\boldsymbol{k}'} - N_{\boldsymbol{k}}(N_{\boldsymbol{k}'} + 1)(N_{\boldsymbol{k}-\boldsymbol{k}'} + 1)]. \tag{6.32}$$

The factor of a half appears for the same reason as in (6.22), and the combinations of occupation numbers in (6.32) are determined by the eigenvalues of the creation and annihilation operators for quasiparticles[‡]. In the classical limit, when $N_{\boldsymbol{k},\boldsymbol{k}',\boldsymbol{k}-\boldsymbol{k}'} \gg 1$, equation (6.30) follows from (6.32) with the probability

$$w(\boldsymbol{k}, \boldsymbol{k}') = 2\pi|V_{\boldsymbol{k}\boldsymbol{k}'}|^2\delta(\omega_{\boldsymbol{k}} - \omega_{\boldsymbol{k}'} - \omega_{\boldsymbol{k}-\boldsymbol{k}'}).$$

This is in direct accordance with the general quantum-mechanical 'golden rule' of calculation of probabilities for quantum processes[§].

Unlike the dynamic equations (6.2) for the amplitudes, the kinetic equation (6.30) does not possess the property of time reversibility. It results in the production of positive entropy in a gas of quasiparticles (see the next problem).

Problem 47

Prove the second law of thermodynamics (the increase of the total entropy) for a gas of quasiparticles whose interaction is described by the kinetic equation (6.30).

In the classical limit of large occupation numbers the entropy density of a

[†] See, Landau L D and Lifshitz E M *Course of Theoretical Physics* vol 10 (Oxford: Pergamon) §2.

[‡] See, Landau L D and Lifshitz E M *Course of Theoretical Physics* vol 3 (Oxford: Pergamon) §64.

[§] See, Landau L D and Lifshitz E M *Course of Theoretical Physics* vol 3 (Oxford: Pergamon) §43.

Bose gas is equal to[†] $S = \int \ln N_k \,\mathrm{d}\boldsymbol{k}$. So the entropy rate of change is

$$\frac{\mathrm{d}S}{\mathrm{d}t} = \int \frac{\mathrm{d}\boldsymbol{k}}{N_k}\frac{\mathrm{d}N_k}{\mathrm{d}t} \qquad (6.33)$$

In the calculation of this integral we need to take into account the change of occupation numbers for all the interacting quasiparticles. Therefore, it follows from (6.33) and (6.30), (6.31) that:

$$\frac{\mathrm{d}S}{\mathrm{d}t} = \int \mathrm{d}\boldsymbol{k}\,\frac{\mathrm{d}N_k}{\mathrm{d}t}\left(\frac{1}{N_k} - \frac{1}{N_{k'}} - \frac{1}{N_{k-k'}}\right)$$

$$= \pi \int \mathrm{d}\boldsymbol{k}\,\mathrm{d}\boldsymbol{k}'\,|V|^2\,\delta(\Delta\omega)\,\frac{(N_{k'}N_{k''} - N_k N_{k'} - N_k N_{k-k'})^2}{N_k N_{k'} N_{k-k'}} \geqslant 0.$$

Thus the total entropy of a gas does not increase only if

$$1/N_k = 1/N_{k'} + 1/N_{k-k'}. \qquad (6.34)$$

It is easy to see that such a relation corresponds to the state of thermodynamical equilibrium in a gas of quasiparticles. Indeed, in the classical limit the Rayleigh–Jeans formula, when $N_k \propto T/\omega_k$, relates to equilibria. In this case the relation (6.34) is satisfied identically because of the resonant condition (6.1) for the frequencies of interacting waves.

It is worth noting that other stationary solutions of the kinetic equation (6.30) may exist, that correspond to a state far from thermodynamic equilibrium. They usually occur in situations when quasiparticles are excited by an external source in some frequency (or wavelength) range, while a dissipative process takes place in another interval. Therefore, such non-equilibrium solutions describe the energy flux over the spectrum of waves from the generation range to the dissipative one[‡].

[†] See, Landau L D and Lifshitz E M *Course of Theoretical Physics* vol 5 (Oxford: Pergamon) part 1, §55.

[‡] See, for instance, Zakharov V E 1983 Kolmogorov spectra in weak turbulence problems *Handbook of Plasma Physics* vol II, eds M N Rosenbluth and R Z Sagdeev (Amsterdam: North-Holland).

PART II

Fluid Mechanics

Chapter 7

Dynamics of Ideal Fluids

Recommended textbooks:
Landau L D and Lifshitz *Course of Theoretical Physics, vol 6, Fluid Mechanics* (Oxford: Pergamon)
Batchelor G K *An Introduction to Fluid Dynamics* (Cambridge: Cambridge University Press)

The motion of an ideal fluid is described by the following equations: the *continuity equation*

$$\partial\rho/\partial t + \operatorname{div}(\rho\boldsymbol{v}) = 0 \tag{7.1}$$

the equation of motion (the Euler equation)

$$\partial\boldsymbol{v}/\partial t + (\boldsymbol{v}\cdot\boldsymbol{\nabla})\boldsymbol{v} = -\boldsymbol{\nabla}p/\rho \tag{7.2}$$

and the adiabatic condition

$$\mathrm{d}S/\mathrm{d}t = \partial S/\partial t + (\boldsymbol{v}\cdot\boldsymbol{\nabla})S = 0 \tag{7.3}$$

where $\rho(\boldsymbol{r}, t)$, $p(\boldsymbol{r}, t)$ and $\boldsymbol{v}(\boldsymbol{r}, t)$ are the density, pressure and velocity of a fluid, respectively, and $S(p, \rho)$ is the entropy per unit mass. For the stationary flow the *Bernoulli equation* follows from (7.2) and (7.3):

$$\boldsymbol{v}\cdot\boldsymbol{\nabla}(\tfrac{1}{2}v^2 + w) = 0 \tag{7.4}$$

that is the constancy of $(v^2/2 + w)$ along a streamline (w is the heat function per unit mass of fluid, or enthalpy). The velocity circulation round a 'closed fluid contour' is conserved for the motion of ideal fluid (the *Kelvin theorem*):

$$\mathrm{d}\oint \boldsymbol{v}\cdot\mathrm{d}\boldsymbol{l}/\mathrm{d}t = 0. \tag{7.5}$$

The equation of motion (7.2) is equivalent to the momentum conservation law in a fluid and may be written in the 'divergent' form:

$$\partial(\rho v_i)/\partial t = -\partial\Pi_{ik}/\partial x_k \tag{7.6}$$

where $\Pi_{ik} = p\delta_{ik} + \rho v_i v_k$ is the *momentum flux density tensor*.

In a similar way the energy conservation law takes the following form:

$$\frac{\partial}{\partial t}\left(\tfrac{1}{2}\rho v^2 + \rho\mathscr{E}\right) = -\operatorname{div}\boldsymbol{q} \tag{7.7}$$

where $\boldsymbol{q} = \rho\boldsymbol{v}(\frac{1}{2}v^2 + w)$ is the *energy flux density vector*, and $\mathscr{E}$ is the internal energy per unit mass.

Problem 48

Derive the equation for the evolution of vorticity $\boldsymbol{\Gamma} \equiv \operatorname{curl}\boldsymbol{v}$ in an ideal fluid. Show that the vector lines of vorticity are 'frozen' in the fluid.

The Euler equation (7.2) may be rewritten in the following form:

$$\frac{\partial \boldsymbol{v}}{\partial t} + (\boldsymbol{v}\cdot\boldsymbol{\nabla})\boldsymbol{v} = -\frac{\boldsymbol{\nabla} p}{\rho} = -\boldsymbol{\nabla} w$$

or

$$\frac{\partial \boldsymbol{v}}{\partial t} - (\boldsymbol{v}\times\operatorname{curl}\boldsymbol{v}) = -\boldsymbol{\nabla}(w + \tfrac{1}{2}v^2).$$

By taking the curl of this equation we have for $\boldsymbol{\Gamma} \equiv \operatorname{curl}\boldsymbol{v}$:

$$\partial\boldsymbol{\Gamma}/\partial t = \operatorname{curl}(\boldsymbol{v}\times\boldsymbol{\Gamma}).$$

We expand the right-hand side, using the fact that $\operatorname{div}\Gamma = 0$:

$$\partial\boldsymbol{\Gamma}/\partial t = (\boldsymbol{\Gamma}\cdot\boldsymbol{\nabla})\boldsymbol{v} - (\boldsymbol{v}\cdot\boldsymbol{\nabla})\boldsymbol{\Gamma} - \boldsymbol{\Gamma}\operatorname{div}\boldsymbol{v}.$$

Substituting from the equation of continuity (7.1)

$$\operatorname{div}\boldsymbol{v} = -\frac{1}{\rho}\frac{\partial\rho}{\partial t} - \frac{1}{\rho}\boldsymbol{v}\cdot\boldsymbol{\nabla}\rho$$

we obtain, after a simple rearrangement of terms,

$$\left(\frac{\partial}{\partial t} + \boldsymbol{v}\cdot\boldsymbol{\nabla}\right)\frac{\boldsymbol{\Gamma}}{\rho} \equiv \frac{\mathrm{d}}{\mathrm{d}t}\left(\frac{\boldsymbol{\Gamma}}{\rho}\right) = \left(\frac{\boldsymbol{\Gamma}}{\rho}\cdot\boldsymbol{\nabla}\right)\boldsymbol{v}.$$

Let us now consider some 'fluid line', that is a line that moves with the fluid particles composing it. Let $\delta\boldsymbol{l}$ be an element of length of this line; we shall determine how $\delta\boldsymbol{l}$ varies with time. If $\boldsymbol{v}$ is the fluid velocity at one end of the element $\delta\boldsymbol{l}$, then the fluid velocity at the other end is $\boldsymbol{v} + (\delta\boldsymbol{l}\cdot\boldsymbol{\nabla})\boldsymbol{v}$. During a time interval $\mathrm{d}t$, the length of $\delta\boldsymbol{l}$ therefore changes by $\mathrm{d}t(\delta\boldsymbol{l}\cdot\boldsymbol{\nabla})\boldsymbol{v}$, that is $\mathrm{d}(\delta\boldsymbol{l})/\mathrm{d}t = (\delta\boldsymbol{l}\cdot\boldsymbol{\nabla})\boldsymbol{v}$. We see that the rates of change of the vectors $\mathrm{d}\boldsymbol{l}$ and $\boldsymbol{\Gamma}/\rho$ are given by identical formulae. Hence it follows that if these vectors are initially in the same direction, they will remain parallel, and their lengths will remain in the same ratio. In other words, if two fluid particles that are initially close to each other are on the same line of the vector field $\mathbf{\Gamma}$ at any time, then they will always be on the same line of the vorticity field. Passing now from particles at an infinitesimal distance apart to those at any distance apart we conclude that every line of the vorticity vector field moves with the fluid particles which lie on it, so the vorticity lines are 'frozen' in the ideal fluid and move with it.

Some simple consequences follow from this fact. Firstly, if the vorticity of the flow is initially equal to zero everywhere, the flow will be potential at any subsequent moment. Secondly, the Kelvin theorem is valid. Indeed, as any closed fluid contour moves about in the course of time, it cuts no lines of vorticity, so the flux of the vorticity vector through any surface spanning the fluid contour does not vary with time. However, according to the Stokes theorem this flux is equal to the velocity circulation round a fluid contour.

Problem 49

Write down the equations for the one-dimensional dynamics of an ideal fluid in terms of Lagrangian variables.

The Lagrangian description is made in terms of the functions $x(x_0, t)\rho(x_0, t)$ and $p(x_0, t)$, where x_0 is the initial position of a fluid element (a layer in the one-dimensional case). The mass conservation condition is now expressed as $\rho\,\mathrm{d}x = \rho_0\,\mathrm{d}x_0$, where $\rho_0(x_0)$ is the initial density, so

$$(\partial x/\partial x_0)_t = \rho_0(x_0)/\rho(x_0, t). \tag{7.8}$$

The equation of motion for a layer of the width $\mathrm{d}x$ is as follows:

$$\rho\,\mathrm{d}x\left(\frac{\partial^2 x}{\partial t^2}\right)_{x_0} = -p(x+\mathrm{d}x) + p(x) + \rho\,\mathrm{d}x f(x_0, t)$$

(f is the external force per unit mass), so

$$\left(\frac{\partial^2 x}{\partial t^2}\right)_{x_0} = -\frac{1}{\rho_0(x_0)}\left(\frac{\partial p}{\partial x_0}\right)_t + f(x_0, t). \tag{7.9}$$

The third equation expresses the conservation of entropy:

$$(\partial S/\partial t)_{x_0} = 0 \tag{7.10}$$

where $S(p, \rho)$ is the entropy per unit mass.

Problem 50

In a uniform medium with the density ρ_0 and zero pressure (dust) the velocity perturbation $v(x) = v_0 \sin(\pi x/l)$ is produced at some moment. Find the resulting distribution of the density.

This problem possesses a simple solution in Lagrangian coordinates since the corresponding equations become linear in this description (unlike in the Euler formalism). Indeed, in the absence of pressure and external force ($p=0, f=0$) it follows from (7.9) that each element of dust continues to move with a constant velocity: $(\partial^2 x/\partial t^2)_{x_0} = 0$, so

$$x = x_0 + v(x_0)t = x_0 + v_0 t \sin(\pi x_0/l).$$

Then we obtain for the density from (7.8):

$$\rho(x_0, t) = \rho_0 \bigg/ \left(\frac{\partial x}{\partial x_0}\right)_t = \rho_0 \left[1 + \frac{\pi v_0 t}{l} \cos\left(\frac{\pi x_0}{l}\right)\right]^{-1}. \tag{7.11}$$

Together with the already known function $x(x_0, t)$ this formula determines, in particular, the density distribution $\rho(x, t)$. The important point here is that the density tends to infinity while a motion proceeds. As is seen from (7.11) this happens when the derivative $(\partial x/\partial x_0)_t$ becomes equal to zero. In our case, according to (7.11), it takes place firstly at the point $x_0 = l$ at the instant $t = t_* = l/\pi v_0$. This phenomenon is called the 'breaking' of a velocity profile, since a graph $v(x, t)$ has an infinite derivative at the point $x_0 = l$ at this moment. For $t > t_*$ the 'intersection of trajectories' occurs, when different elements of a fluid with different initial positions x_0 have the same coordinate x. Therefore, the relation $x(x_0, t)$ is no longer one to one, so the Lagrangian description fails.

Problem 51

Describe the expansion of a uniformly charged spherical cloud of dust, which is initially at rest.

Let us denote the initial density of the dust as ρ_0 and the initial charge density at $\alpha\rho_0$ (so α is the charge per unit mass). In the Lagrangian description we are interested in the functions $r(r_0, t)$ and $v(r_0, t)$. If a profile 'breaking' is absent (this will be confirmed by what follows) the net charge inside each expanding spherical surface remains constant:

$$q(r) = q(r_0) = \tfrac{4}{3}\alpha\rho_0\pi r_0^3.$$

Then the equation of motion can be written in the form:

$$(\partial^2 r/\partial t^2)_{r_0} = f(r_0, t) = \alpha E(r_0, t) = \alpha^2 \rho_0 (\tfrac{4}{3}\pi r_0^3 r^{-2}).$$

By integrating this equation with the initial conditions $r(0) = r_0$ and $\dot{r}(0) = 0$, we get

$$\omega_0 t = \int_1^{r/r_0} \mathrm{d}x (1 - 1/x)^{-1/2} \tag{7.12}$$

where the quantity ω_0 possesses the dimensional representation of a frequency and equals $(8\pi\alpha^2\rho_0/3)^{1/2}$. This equation determines (in a non-direct form) the relation $r(r_0, t)$. The latter may be represented as $r = r_0\mathscr{F}(\omega_0 t)$. The lack of profile 'breaking' is now evident. In addition this means that a uniform expansion of a dust occurs, so its density inside the cloud remains homogeneous: $\rho(t) = \rho_0\mathscr{F}^{-3}(\omega_0 t)$. At $\omega_0 t \gg 1$ the function $\mathscr{F}$ is simply $\mathscr{F}(\omega_0 t) \approx \omega_0 t$. This means that when the dust cloud has greatly expanded, the electric fields become negligibly weak, so that each spherical layer continues expanding with almost constant velocity.

Problem 52

Find the criterion for a profile 'breaking' in the spherical dust cloud considered in the previous problem for a non-homogeneous initial density distribution.

From the equation of motion for a charged dust cloud the following generalization of the relation (7.12) takes place for a non-uniform initial density:

$$\int_1^{r/r_0} \mathrm{d}x (1 - x^{-1})^{-1/2} = \omega(r_0) t$$

$$\omega(r_0) = \left(8\pi\alpha^2 \int_0^{r_0} \rho_0(x) x^2 \, \mathrm{d}x \right)^{1/2} r_0^{-3/2}. \tag{7.13}$$

The absence of 'breaking' means that the derivative $(\partial r/\partial r_0)_t$ is positive. From (7.13) we get for the latter:

$$\left(\frac{\partial r}{\partial r_0}\right)_t = t r_0 \frac{\mathrm{d}\omega}{\mathrm{d}r_0} \left(1 - \frac{r_0}{r}\right)^{1/2} + \frac{r}{r_0}.$$

By eliminating t with the help of (7.13) we find the following condition for the absence of 'breaking':

$$\frac{r_0}{\omega}\frac{\mathrm{d}\omega}{\mathrm{d}r_0} > -\frac{x/(1-x^{-1})^{1/2}}{\int_1^x \mathrm{d}z(1-z^{-1})^{-1/2}} \equiv -p(x) \qquad x = \frac{r}{r_0}. \tag{7.14}$$

This inequality must be satisfied throughout the whole interval $1 \leqslant x \leqslant +\infty$. So we need to consider the minimum of the function $p(x)$ in this interval. After a simple integration we have for $p(x)$:

$$p(x) = \left(1 - \frac{1 + (1-x^{-1})^{1/2}\ln[x^{1/2} + (x-1)^{1/2}]}{x}\right)^{-1}.$$

This is a positive function, and $p(x) \to +\infty$ for $x \to 1$, and $p(x) \to 1$ at $x \to +\infty$. It may be found numerically that $\min p(x) \approx 0.93$ which is achieved at $x \approx 8$. So the condition for the absence of profile 'breaking' is, according to (7.14),

$$\rho_0(r_0)r_0^3 > 1.14 \int_0^{r_0} \rho_0(x)x^2\,\mathrm{d}x.$$

This inequality must be valid for each r_0 less than or equal to the initial radius of a cloud R_0. For instance, if the density has a power law distribution: $\rho_0(r_0) \propto r_0^n$, profile 'breaking' occurs for $n \leqslant -1.86$.

Problem 53

In a cold plasma containing infinitely heavy ions and electrons with mass m, charge e and density n_0, an electron velocity distribution $v(x) = v_0 \sin(\pi x/L)$ is produced at some moment. Determine the critical value of the velocity amplitude v_0 that provokes the electron stream 'breaking' when it is exceeded.

The only force acting upon the electrons is an electrical one, so, according to equation (7.9),

$$(\partial^2 x/\partial t^2)_{x_0} = eE/m$$

where E is the electric field created by the charge separation during the motion of the electrons. Because of the fixed neutralizing background of the plasma ions the displacement of the electron layer with initial coordinate x_0 to the new position with coordinate x will lead to the formation of a redundant charge (per unit area) $\sigma = n_0 e(x_0 - x)$ on one side of the layer and to charge of the opposite sign on the other side (we suppose here that there is no 'intersection' of electron trajectories). As a result the electric field $E = 4\pi\sigma = 4\pi n_0 e(x_0 - x)$ will appear and we can easily integrate the equation for the electron motion:

$$(\partial^2 x/\partial t^2)_{x_0} = eE/m = -\omega_{pe}^2(x - x_0) \qquad \omega_{pe}^2 = 4\pi n_0 e^2/m$$
$$x - x_0 = A \sin \omega_{pe} t + B \cos \omega_{pe} t. \tag{7.15}$$

The constants A and B have to be determined from the initial conditions, namely:

$$(x - x_0)|_{t=0} = 0 \qquad (\partial x/\partial t)_{t=0} = v_0 \sin(\pi x_0/L).$$

So we obtain from (7.15) that

$$x(x_0, t) = x_0 + \frac{v_0}{\omega_{pe}} \sin\left(\frac{\pi x_0}{L}\right) \sin \omega_{pe} t.$$

The compression rate is then

$$\left(\frac{\partial x}{\partial x_0}\right)_t^{-1} = \left[1 + \frac{\pi v_0}{\omega_{pe} L} \cos\left(\frac{\pi x_0}{L}\right) \sin \omega_{pe} t\right]^{-1}.$$

It is seen from this expression that the most 'dangerous' place for 'breaking' is the layer with maximum gradient: $x_0 = L$. At this point the 'intersection' of the electron trajectories takes place if $\pi v_0/\omega_{pe} L > 1$, so the critical amplitude is

$$v_{cr} = \omega_{pe} L/\pi.$$

Problem 54

The stationary flow of an ideal incompressible fluid with density ρ is rotated by an angle α by a tube of variable cross section and then ejected into a vacuum (figure 7.1). Find the force acting on the bend in the tube. Consider the flow as uniform at the cross sections S_0 and S_1: the inlet velocity is equal to v_0.

For a stationary flow the net momentum flux over any closed surface must be equal to zero: $\int \Pi_{ik}\, dS_k = 0$, where Π_{ik} is the momentum flux density tensor given by the expression (7.6): $\Pi_{ik} = p\delta_{ik} + \rho v_i v_k$. Let us choose the closed surface formed by the lateral surface of the bend in the tube and by the inlet and outlet cross sections (figure 7.1). Then the lack of a net momentum flux may be written as follows:

$$n_{0k} S_0 (p_0 \delta_{ik} + \rho v_{0i} v_{0k}) + n_{1k} S_1 \rho v_{1i} v_{1k} + \int p\delta_{ik}\, dS_k = 0. \tag{7.16}$$

Here $\boldsymbol{n}_0$ and $\boldsymbol{n}_1$ are the unit vectors normal to the inlet and outlet cross sections, $\boldsymbol{v}_1$ is the outlet velocity, and p_0 is the inlet pressure of the fluid. The integral in (7.16) is carried out over the lateral surface of the tube (the

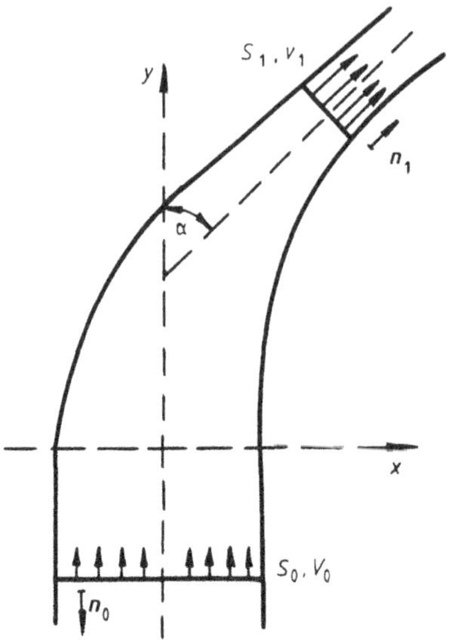

Figure 7.1. Rotation of a flow by a pipe.

term $\rho v_i v_k$ does not contribute there, since the fluid velocity is tangential at this surface: $v_k \, dS_k = 0$). The force of interest is equal to this integral, so

$$\mathscr{F}_i = \int p \, dS_i = -S_0 p_0 n_{0i} + \rho v_0 S_0 v_{0i} - \rho v_1 S_1 v_{1i}.$$

The outlet velocity is determined from the constancy of a fluid discharge in a tube: $\rho v_0 S_0 = \rho v_1 S_1$, so $v_1 = v_0 S_0 / S_1$. The inlet pressure p_0 needed to push a fluid through a tube can be obtained from the Bernoulli equation. For an incompressible fluid this takes the form: $p + \frac{1}{2}\rho v^2 = \text{const}$, therefore,

$$p_0 = \tfrac{1}{2}\rho v_0^2[(S_0/S_1)^2 - 1]$$

(we consider that $S_0 > S_1$, so $p_0 > 0$). By introducing the coordinate system shown in figure 7.1, the final result may be written as

$$\mathscr{F}_x = -\frac{\rho v_0^2 S_0^2}{S_1} \sin\alpha$$

$$\mathscr{F}_y = \rho v_0^2 S_0 \left[\frac{1}{2}\left(\frac{S_0^2}{S_1^2} + 1 \right) - \frac{S_0}{S_1} \cos\alpha \right].$$

Problem 55

Derive the dispersion relation for the surface waves propagating along the horizontal boundary of two incompressible fluids with the densities ρ_1 and ρ_2 in the gravitational field g. The surface-tension coefficient is equal to α.

For a wave with a small amplitude the motion of fluids may be considered as a potential[†], so the linearized equations of motion and continuity take the form:

$$\rho\, \partial \boldsymbol{v}/\partial t = -\nabla \delta p \qquad \operatorname{div} \boldsymbol{v} = 0 \qquad \boldsymbol{v} = -\nabla \varphi \tag{7.17}$$

where δp is the pressure perturbation in a wave. In the unperturbed state the pressure distribution in the fluids 1 and 2 is $p_1 = p_0 - \rho_1 g z$; $p_2 = p_0 - \rho_2 g z$, where p_0 is the pressure at $z = 0$ (figure 7.2). If a wave propagates along

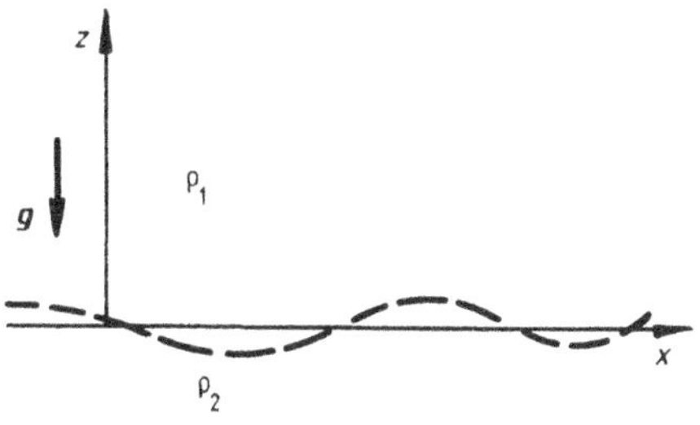

Figure 7.2. Surface waves in a fluid.

the x-axis, the perturbed quantities may be written as follows:

$$\varphi = f(z)\,\mathrm{e}^{\mathrm{i}(kx-\omega t)} \qquad \delta p = \psi(z)\,\mathrm{e}^{\mathrm{i}(kx-\omega t)} \qquad \xi = \xi_0\,\mathrm{e}^{\mathrm{i}(kx-\omega t)}$$

where ξ is a vertical displacement of the fluid boundary. We shall determine the functions $f(z)$ and $\psi(z)$ separately in the upper and lower fluids. The velocity potential φ satisfies, according to (7.17), the equation $\nabla^2\varphi = 0$. Since we are interested in a surface wave, all the perturbations have to decrease as we go to the interior of the fluids, so

$$\varphi_1 = A\,\mathrm{e}^{-kz}\,\mathrm{e}^{\mathrm{i}(kx-\omega t)} \qquad \varphi_2 = B\,\mathrm{e}^{kz}\,\mathrm{e}^{\mathrm{i}(kx-\omega t)}.$$

The constants A and B are not independent of each other but are linked by the continuity of the normal component of the fluid velocity at the boundary. Since φ is a small perturbation itself, this boundary condition may be calculated at the unperturbed surface $z = 0$. Thus

$$-\partial\varphi_1/\partial z|_{z=0} = -\partial\varphi_2/\partial z|_{z=0} = \partial\xi/\partial t$$

so $B = -A = \mathrm{i}\omega\xi_0/k$. By substituting this into the linearized equation of

[†] See, Landau L D and Lifshitz E M *Course of Theoretical Physics* vol 6 (Oxford: Pergamon) §9.

motion (7.17) we obtain for the pressure perturbations in the fluids:

$$\delta p_1 = -\mathrm{i}\omega\rho_1\varphi_1 = \frac{-\omega^2\rho_1\xi_0\,\mathrm{e}^{-kz}}{k}\,\mathrm{e}^{\mathrm{i}(kx-\omega t)}$$

$$\delta p_2 = -\mathrm{i}\omega\rho_2\varphi_2 = \frac{\omega^2\rho_2\xi_0\,\mathrm{e}^{kz}}{k}\,\mathrm{e}^{\mathrm{i}(kx-\omega t)}.$$

The dispersion relation $\omega(k)$ may be derived now from the boundary condition for the pressure balance. It follows from the Laplace formula that

$$(p_1 + \delta p_1)|_{z=\xi} - (p_2 + \delta p_2)|_{z=\xi} = \alpha/R$$

where R is the radius of curvature of the boundary surface taken with a proper sign (R is positive if the surface $z = \xi$ is convex downwards). For a small displacement, when $k\xi_0 \ll 1$,

$$R^{-1} \approx \partial^2\xi/\partial x^2 = -k^2\xi_0\,\mathrm{e}^{\mathrm{i}(kx-\omega t)}.$$

Then only taking into account the terms of the first order on perturbations, we have:

$$\xi_0\,\mathrm{e}^{\mathrm{i}(kx-\omega t)}\left(-\rho_1 g - \frac{\omega^2\rho_1}{k} + \rho_2 g - \frac{\omega^2\rho_2}{k}\right) = -\alpha k^2\xi_0\,\mathrm{e}^{\mathrm{i}(kx-\omega t)}$$

and finally

$$\omega(k) = \left(\frac{gk(\rho_2 - \rho_1) + \alpha k^3}{\rho_1 + \rho_2}\right)^{1/2}. \tag{7.18}$$

Let us now consider some particular cases.

(a) The case where $\rho_1 = 0$ and $\rho_2 = \rho$ is the free boundary of a heavy fluid. In this case $\omega(k) = (gk + \alpha k^3/\rho)^{1/2}$. For a long wavelength, when $k \ll (\rho g/\alpha)^{1/2}$, $\omega \approx (gk)^{1/2}$, and this is a gravitational surface wave. In another limit, for short wavelengths ($k \gg (\rho g/\alpha)^{1/2}$), $\omega \approx (\alpha k^3/\rho)^{1/2}$. Such a wave determined by the surface pressure is called a capillary wave. If $\rho_1 \neq 0$, but $\rho_1 < \rho_2$ (a light fluid above a heavy one) the spectrum of the surface waves remains qualitatively the same.

(b) The case where $\rho_1 \neq 0$ and $\rho_1 > \rho_2$ (a heavy fluid above a light one). In this situation the boundary surface becomes unstable. If the surface pressure is absent the instability growth rate is

$$\gamma = \mathrm{i}\omega = [gk(\rho_1 - \rho_2)/(\rho_1 + \rho_2)]^{1/2}.$$

This is the Rayleigh–Taylor instability. The capillarity leads to stabilization of the short-wavelength perturbations with $k > k_0 = [g(\rho_1 - \rho_2)/\alpha]^{1/2}$, but instability is still present for the long-wavelength ones.

Problem 56

Investigate the Rayleigh–Taylor instability for a finite width of the interface between the light and heavy incompressible fluids, when the density distribution is

$$\rho(z) = \begin{cases} \rho_2 & z \leqslant 0 \\ \rho_2\, e^{z/h} & 0 \leqslant z \leqslant h \\ \rho_1 = e\,\rho_2 & z \geqslant h. \end{cases} \tag{7.19}$$

In an incompressible fluid the density of any fluid element remains constant as it moves about in space, so a linearized equation for the density perturbation at a fixed point in space, $\delta\rho$, is

$$d(\rho + \delta\rho)/dt \approx \partial\delta\rho/\partial t + \boldsymbol{v}\cdot\nabla\rho = 0.$$

Together with the linearized continuity equation

$$\partial\delta\rho/\partial t + \mathrm{div}(\rho\boldsymbol{v}) = 0$$

this gives that

$$\mathrm{div}\,\boldsymbol{v} = 0. \tag{7.20}$$

The linearized equation of motion is

$$\rho\,\partial\boldsymbol{v}/\partial t = -\nabla\delta p + \boldsymbol{g}\delta\rho \tag{7.21}$$

where δp is a pressure perturbation. Because of the symmetry in the horizontal $(x-y)$ plane it is always possible to write the perturbations in a form proportional to $f(z)\,e^{ikx+\gamma t}$. Then it follows from (7.20) that

$$ikv_x + dv_z/dz = 0$$

that is

$$v_x = (i/k)\,dv_z/dz$$

and from the continuity equation that

$$\gamma\delta\rho + v_z\,d\rho/dz = 0$$

that is

$$\delta\rho = -\frac{v_z}{\gamma}\,d\rho/dz.$$

From the x-component of (7.21) we find that

$$\gamma\rho v_x = -ik\delta p$$

that is

$$\delta p = \mathrm{i}\gamma \rho v_x / k = -\frac{\gamma \rho}{k^2} \frac{\mathrm{d}v_z}{\mathrm{d}z}.$$

By substituting these expressions for $\delta\rho$ and δp we obtain the following equation for v_z from the z-component of (7.21):

$$\rho^{-1} \frac{\mathrm{d}}{\mathrm{d}z}\left(\rho \frac{\mathrm{d}v_z}{\mathrm{d}z}\right) + \frac{gk^2}{\gamma^2 \rho} \frac{\mathrm{d}\rho}{\mathrm{d}z} v_z - k^2 v_z = 0$$

or a more convenient equation for the new variable $q = \rho^{1/2} v$:

$$\frac{\mathrm{d}^2 q}{\mathrm{d}z^2} - k^2 \left(1 - \frac{g}{\gamma^2 \rho} \frac{\mathrm{d}\rho}{\mathrm{d}z} + \frac{1}{k^2 \rho^{1/2}} \frac{\mathrm{d}^2 \rho^{1/2}}{\mathrm{d}z^2}\right) q = 0. \qquad (7.22)$$

For the density profile (7.19) its solution is as follows:

$$q(z) = \begin{cases} A_1 \, \mathrm{e}^{kz} & z \leqslant 0 \\ A_2 \, \mathrm{e}^{\kappa z} + B_2 \, \mathrm{e}^{-\kappa z} & 0 \leqslant z \leqslant h \cdot \\ A_3 \, \mathrm{e}^{-kz} & z \geqslant \mathrm{h} \end{cases} \qquad (7.23)$$

where

$$\kappa^2 = k^2 \left(1 - \frac{g}{\gamma^2 h} + \frac{1}{4k^2 h^2}\right)$$

and only solutions decreasing inside the uniform fluids are retained. The constants A_2, B_2 and A_3 have to be found from the matching conditions at $z = 0$ and $z = h$. The latter consists in the continuity of q at these points and an appropriate change of its derivative, $\mathrm{d}q/\mathrm{d}z$, there. Indeed, let us integrate equation (7.22) over an infinitesimal interval around $z = 0$, then

$$\int_{0-\varepsilon}^{0+\varepsilon} \frac{\mathrm{d}^2 q}{\mathrm{d}z^2} \mathrm{d}z - \int_{0-\varepsilon}^{0+\varepsilon} \frac{1}{\rho^{1/2}} \frac{\mathrm{d}^2 \rho^{1/2}}{\mathrm{d}z^2} q \, \mathrm{d}z = 0$$

or

$$\left.\frac{\mathrm{d}q}{\mathrm{d}z}\right|_{0-\varepsilon} - \left.\frac{\mathrm{d}q}{\mathrm{d}z}\right|_{0+\varepsilon} = \frac{q(0)}{[\rho(0)]^{1/2}} \left(\left.\frac{\mathrm{d}\rho^{1/2}}{\mathrm{d}z}\right|_{0+\varepsilon} - \left.\frac{\mathrm{d}\rho^{1/2}}{\mathrm{d}z}\right|_{0-\varepsilon} \right) = \frac{q(0)}{2h}.$$

By performing a similar procedure at $z = h$ we find from (7.23) that

$$A_2 = A_1 (k + \kappa + 1/2h)/2\kappa$$

$$B_2 = \mathrm{A}_1 (\kappa - k - 1/2h)/2\kappa$$

$$A_3 = \mathrm{e}^{kh} (A_2 \, \mathrm{e}^{\kappa h} + B_2 \, \mathrm{e}^{-\kappa h})$$

and the dispersion equation, which determines the growth rate γ:

$$e^{2\kappa h} = [(4h^2)^{-1} - (k-\kappa)^2]/[(4h^2)^{-1} - (k+\kappa)^2]$$

or, after the simple transformation,

$$(\tanh \kappa h)/\kappa h = 2kh/(\tfrac{1}{4} - k^2h^2 - \kappa^2h^2). \tag{7.24}$$

Two different cases depending on the sign of κ^2 must be considered. Let us start with $\kappa^2 > 0$. Then for positive κ the left-hand side of equation (7.24) is a decreasing function of κh, while the right-hand side is an increasing one. So a solution of (7.24) exists if at $\kappa h = 0$ the left-hand side is larger than the right-hand side:

$$1 \geqslant 2kh/(\tfrac{1}{4} - k^2h^2)$$

or $kh \leqslant (\sqrt{5}/2 - 1)$. It is seen from (7.24) that for long-wavelength perturbations, with $kh \ll 1$, the root of this equation must be very close to a half, so we put $\kappa h \approx \frac{1}{2} - \varepsilon$, where $\varepsilon \ll 1$. By a simple expansion in (7.24) we obtain that $\varepsilon \approx kh/(\tanh \frac{1}{2})$. Now taking into account the definition of κ in (7.23) we find the instability growth rate:

$$\gamma^2 \approx gk \tanh \tfrac{1}{2} = gk\frac{e-1}{e+1} = gk\frac{\rho_1 - \rho_2}{\rho_1 + \rho_2} \tag{7.25}$$

the result, which was obtained above in the previous problem from the sharp-boundary model. The growth rate increases with larger kh and becomes of the order of $(g/h)^{1/2}$ for $kh \sim 1$.

In the case when $\kappa^2 < 0$, equation (7.24) may be rewritten in the following form:

$$(\tan \alpha h)/\alpha h = 2kh/(\tfrac{1}{4} - k^2h^2 + \alpha^2h^2) \tag{7.26}$$

where $\alpha = i\kappa$. It is convenient to solve this equation graphically by plotting its left-hand and right-hand sides. It is then seen that an infinite number of roots exist for any given value of kh. If $kh \ll 1$, these roots are very close to the nulls of $\tan \alpha h$ so $\alpha h \approx n\pi$, $n = 1, 2, \ldots$. The corresponding values of the growth rate are $\gamma \sim (gk)^{1/2}(kh)^{1/2}$, that is significantly less than (7.25) (and γ becomes even smaller for large n). For $kh \sim 1$ the maximum growth rate is of the order of $(g/h)^{1/2}$ and remains of the same order for $kh \gg 1$.

Finally, the following conclusions hold true. For a long-wavelength perturbation the sharp-boundary model is adequate; it gives the correct expression for the growth rate of the most unstable modes: $\gamma \sim (gk)^{1/2}$. When the wavelength of a perturbation $\lambda \sim k^{-1}$ is comparable with the width of the transition layer, the specific structure of the density profile becomes important. The short-wavelength perturbations ($\lambda \ll h$) are localized inside the transition layer, and their growth rate is saturated at a value of the order of $(g/h)^{1/2}$.

Problem 57

The quasimonochromatic wave packet of gravitational waves, containing N, $(N \gg 1)$, crests and wells, propagates along the fluid surface (figure 7.3). How many 'up and down' motions will be undergone by a light float while the packet passes?

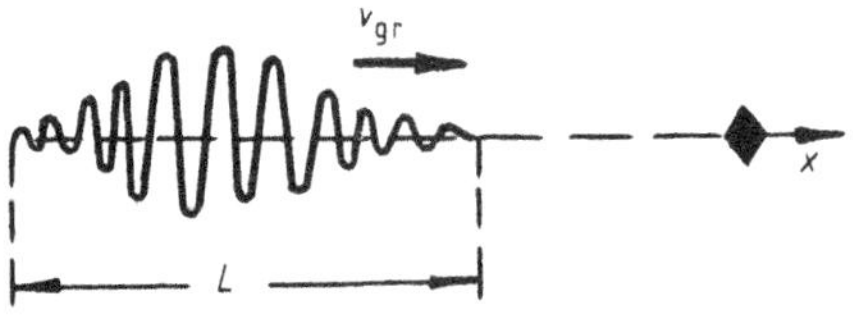

Figure 7.3. Propagation of a wave packet of gravitational waves.

The perturbation of a fluid surface for such a packet may be written in the form

$$\xi(x,t) = \mathscr{F}(x - v_g t)\cos(kx - \omega t)$$

where $\mathscr{F}(x,t)$ is the packet envelope (see problem 30). Since the length of the packet is equal to $L = N\lambda = 2\pi N/k$, and it propagates with a group velocity v_g, a float will be moved by the packet during an interval of time $\tau = L/v_g = 2\pi N/kv_g$. For a quasimonochromatic packet the period of the fluid's surface oscillation is $T = 2\pi/\omega(k)$, so the number of undulations of the float is

$$n = \tau/T = 2\pi N\omega(k)/2\pi k v_g = N\omega(k)/kv_g = Nv_{ph}/v_g.$$

The dispersion relation for a gravitational wave is $\omega(k) = (gk)^{1/2}$, so its phase velocity, v_{ph}, is twice as large as its group velocity v_g. Thus $2N$ motions will be undergone by the float.

Problem 58

When a horizontal surface of water is perturbed by an instant point source (for instance, by a falling stone), circles of gravitational water waves start to propagate outwards from this point. As a result the vertical displacement of a surface represents a set of alternating crests and wells inside an expanding circle of radius $R(t)$. Estimate the value of $R(t)$.

Crests and wells on a water surface represent the lines (circles in this particular case) of constant phases for surface oscillations, and the phase difference between adjacent crests and wells is equal to π. Let us consider the phase of the gravitational wave as a function of the distance from the source at a given moment. Perturbations produced by a source will propagate with the group velocity $v_g = \partial\omega/\partial k$, so at the moment t and at the distance r it will be a wave with such a wave number k, that $v_g(k) = r/t$. Since $\omega(k) = (gk)^{1/2}$, $v_g(k) = \frac{1}{2}(g/k)^{1/2}$ and $k(r,t) = gt^2/4r^2$. The phase of the oscillation is $\varphi(r,t) = \omega t - kr = gt^2/4r$. It is seen from this expression for φ that at a given moment t the phase changes sharply at small distances, so there are many crests and wells very close together there. As r increases the phase varies more and more smoothly. For $r \to \infty$, $\varphi \to 0$ and for $r \sim r_1 \sim gt^2/4\pi$ the corresponding phase is $\varphi(r_1) \sim \pi$. This means that the difference of the phase for a circle with $r > r_1$ and one with r up to infinity is less than π. So there are no alternating crests and wells at these distances. Thus, $R(t) \sim r_1 \sim gt^2/4\pi$.

Problem 59

Determine the wake pattern formed on the fluid surface by gravitational waves behind the point source moving with a constant velocity.

The pattern formed behind a moving source represents a superposition of perturbations, produced by the source at each point of its trajectory, and then propagating along the surface. As a result the net vertical displacement of the fluid surface will be appreciable only at those points where these separate perturbations interfere constructively. Therefore, the lines of crests (or wells) must be the contours of constant phase. Let us consider the elementary act of a superposition of two adjacent circular waves with the same phase emitted with the time interval Δt.

Let a source be at the origin of the coordinate system at the initial moment ($t = 0$) with velocity equal to $-v_0$ directed along the x-axis. Then the centre of the second wave will be displaced to the left by a distance $v_0\Delta t$ (figure 7.4), and its radius will be less than the radius of the first circle by the value $v_{ph}\Delta t$ (since a line of the constant phase propagates with the phase velocity v_{ph}). Superposition of waves with the same phases takes place on the tangent line to these circles, so the wave vector $\boldsymbol{k}$ is directed here in such a way that $\cos\theta = -k_x/k = v_{ph}/v_0$, thus the usual Cherenkov resonance condition is satisfied (see problem 32). The only complication to this simple picture originates from the dispersion of the gravitational waves (their phase velocity depends on the wavelength). Therefore, the angle θ will be varied along a crest line, since the waves of different wavelength will interfere at different points of this line. Indeed, at the instant t the radius equal to r will correspond to the wave for which $r = v_g t = \frac{1}{2}v_{ph}t = \frac{1}{2}v_0 t\cos\theta$. In our coordinate system

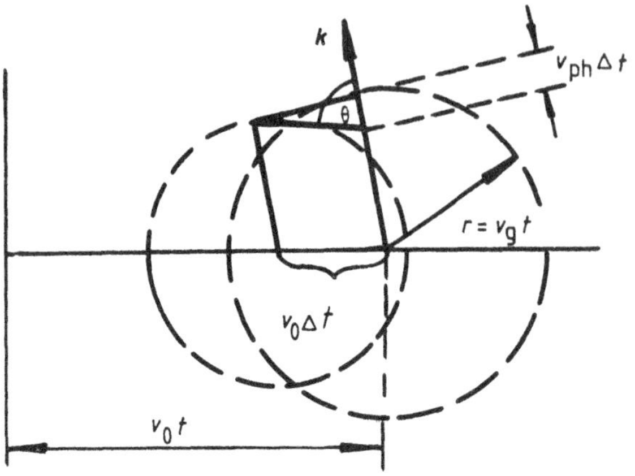

Figure 7.4. Superposition of two circular perturbations.

$y = r \sin \theta$, $x = v_0 t - r \cos \theta$, and the phase, according to the result obtained in the previous problem, is equal to $\varphi = gt^2/4r$. So we may express the coordinates of the point, where the interference takes place, in terms of the wave phase φ and the angle θ (the angle between the wave vector at this point and the x-axis):

$$\begin{aligned} x &= v_0 t - r \cos \theta = (v_0^2 \varphi / g) \cos \theta (2 - \cos^2 \theta) \\ y &= r \sin \theta = (v_0^2 \varphi / g) \sin \theta \cos \theta. \end{aligned} \tag{7.27}$$

Since the phase is constant along any crest (or well) line, the relations (7.27) determine these lines in parametric form. The result is shown in figure 7.5: the angle θ varies from 0 to $\pi/2$. The wake just behind a source corresponds to angles close to $\pi/2$, that is to waves with a small phase velocity ($v_{ph} \ll v_0$, short-wavelength perturbations). On the contrary, $\theta = 0$ at the point A, so here $v_{ph} = v_0$ and for this wave $k = g/v_0^2$. All the crest (and well) lines are geometrically similar, and differ only by changing the phase φ in (7.27) by 2π. All these lines are located inside some angle α_0 (see

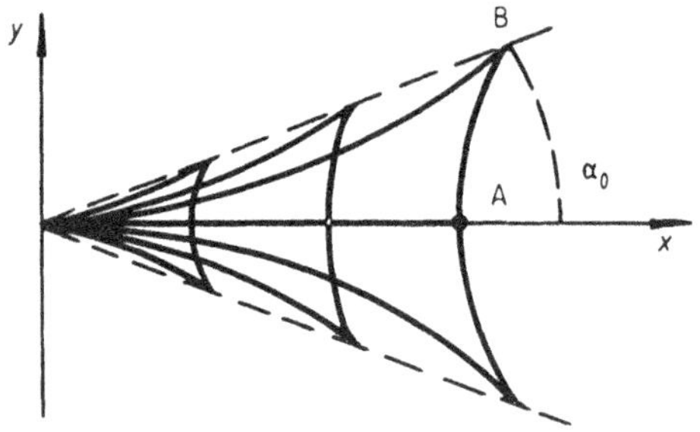

Figure 7.5. The wake pattern of gravitational waves.

figure 7.5). Its value may be found from the condition that x and y in (7.27) take their maximums at the point B. Then it follows from (7.27) that $\sin\theta_B = 1/\sqrt{3}$, so

$$\tan\alpha_0 = y_{max}/x_{max} = \sin\theta_B\cos\theta_B/(2-\cos^2\theta_B) = 1/(2\sqrt{2}).$$

Thus $\alpha_0 \approx 19.5°$.

Problem 60

Investigate the stability of a plane horizontal boundary between two ideal incompressible fluids in the gravitational field $\boldsymbol{g}$, if the upper fluid is considerably lighter than the lower one ($\rho_1 \ll \rho_2$) and is moving horizontally with a velocity equal to V_0 (a 'wind over the ocean'). The surface tension coefficient of this boundary is α.

Let us start with the sharp-boundary model (a tangential discontinuity). Considering the motion in both fluids as a potential, we obtain in the linear approximation the following equations for $z \geqslant 0$ (see also problem 55):

$$\varphi_1 = A\,e^{-kz}\,e^{i(kx-\omega t)}$$

$$\rho_1(\partial\boldsymbol{v}_1/\partial t + V_0\,\partial\boldsymbol{v}_1/\partial x) = -\nabla\delta p_1$$

that is, $\delta p_1 = -i(\omega - kV_0)\rho_1\varphi_1$ (the most unstable perturbations, propagating along the wind, are considered). For $z < 0$ the solution remains as before:

$$\varphi_2 = B\,e^{kz}\,e^{i(kx-\omega t)}$$

that is $\delta p_2 = -i\omega\rho_2\varphi_2$. If a vertical displacement of the boundary surface is $\xi(x,t) = \xi_0\,e^{i(kx-\omega t)}$, the continuity of the normal components of the fluid velocities at the boundary gives the following relation:

$$-\left.\frac{\partial\varphi_1}{\partial z}\right|_{z=\xi} - V_0\frac{\partial\xi}{\partial x} = -\left.\frac{\partial\varphi_2}{\partial z}\right|_{z=0} = \frac{\partial\xi}{\partial t}$$

so that $B = i\omega\xi_0/k$ and $A = -i(\omega - kV_0)\xi_0/k$. Then taking into account the Laplace formula

$$(p_1+\delta p_1)|_{z=\varepsilon} - (p_2+\delta p_2)|_{z=\xi} = \alpha/R$$

we obtain the following dispersion relation for the perturbations:

$$\omega^2 - 2\varepsilon kV_0\omega - (gk + \alpha k^3/\rho_2 - \varepsilon k^2V_0^2) = 0$$

where $\varepsilon \equiv \rho_1/\rho_2 \ll 1$. Its solution is

$$\omega_{1,2} \approx \varepsilon kV_0 \pm (gk + \alpha k^3/\rho_2 - \varepsilon k^2V_0^2)^{1/2}. \tag{7.28}$$

The instability arises when $\mathrm{Im}\,\omega < 0$. Therefore, a perturbation with a given wave number k becomes unstable if

$$V_0 > \varepsilon^{-1/2}\left(\frac{g}{k} + \frac{\alpha k}{\rho_2}\right)^{1/2}.$$

The right-hand side of this inequality depends on k and possesses its minimum at $k = k_0 = (g\rho_2/\alpha)^{1/2}$. So the perturbation with this wave number is the most unstable, and the corresponding critical velocity $V_{cr} = \varepsilon^{-1/2}U_0$, where $U_0 = (4\alpha g/\rho_2)^{1/2}$.

This instability (commonly called the Kelvin–Helmholtz instability) requires a strong coupling between the two fluids since it occurs only when the dispersion equation (7.28) becomes considerably different from that for gravitational-capillary waves. In this sense it is a *gross hydrodynamical instability* and its critical velocity contains the large factor $\varepsilon^{-1/2}$, so it tends to infinity with $\rho_1 \to 0$. This result suffers a qualitative change if even an infinitesimal viscosity of the fluids is taken into account. In this case a viscous boundary layer with a shear flow is formed between the fluids, and a *fine resonant instability* is possible when the wind velocity at some elevation becomes equal to the phase velocity of a gravitational-capillary wave. Thus the critical wind velocity for the wave excitation is reduced to the minimum phase velocity of the gravitational-capillary waves, which is just equal to U_0. This situation may be designated as a weak coupling of the fluids, since the wind has no significant influence on the dispersion of the surface waves. So let us consider this case in detail. Owing to the large difference in densities we may neglect any mean motion of the water that may be induced by the mean air flow. Thus, in the unperturbed state the water is at rest in the domain $z < 0$ and the air has some shear velocity distribution along the x-axis, $V_0(z)$ with $V_0(0) = 0$. Assuming the viscosity is negligible the following linearized equations for the velocity perturbation $\boldsymbol{v}_1$ and for the pressure perturbation δp_1 in air are valid:

$$\rho_1[\partial \boldsymbol{v}_1/\partial t + (\boldsymbol{V}_0\cdot\nabla)\boldsymbol{v}_1 + (\boldsymbol{v}_1\cdot\nabla)\boldsymbol{V}_0] = -\nabla\delta p_1 \qquad (7.29a)$$

$$\operatorname{div}\boldsymbol{v}_1 = 0. \qquad (7.29b)$$

Introducing the stream function $\psi_1(x, z)$, such that $v_{1x} = \partial\psi_1/\partial z$, $v_{1z} = -\partial\psi_1/\partial x$, and equation (7.29$b$) is satisfied identically, we obtain from equation (7.29a) for the perturbations proportional to $\exp(\mathrm{i}kx - \mathrm{i}\omega t)$:

$$\rho_1\left[\left(V_0 - \frac{\omega}{k}\right)\frac{\partial\psi_1}{\partial z} - \frac{\mathrm{d}V_0}{\mathrm{d}z}\psi_1\right] = -\delta p_1$$
$$\rho_1 k^2(V_0 - \omega/k)\psi_1 = -\partial\delta p_1/\partial z \qquad (7.30)$$

or, by elimination of δp_1,

$$\frac{\mathrm{d}^2\psi_1}{\mathrm{d}z^2} - k^2\psi_1 + \frac{kV_0''}{\omega - kV_0}\psi_1 = 0 \qquad (7.31)$$

The corresponding equation for the stream function in water, ψ_2, is

$$\frac{d^2\psi_2}{dz^2} - k^2\psi_2 = 0$$

with the solution that tends to zero at infinity,

$$\psi_2 = A\, e^{kz}\, e^{i(kx-\omega t)}. \tag{7.32}$$

The boundary conditions at the air–water interface are as follows:

$$\psi_1(0) = \psi_2(0) = (\omega/k)\xi_0 \tag{7.33}$$

(ξ_0 is the amplitude of the boundary displacement). Since according to (7.30) and (7.32) the pressure perturbation in the water is

$$\delta p_2 = \rho_2 \frac{\omega}{k}\frac{d\psi_2}{dz} = \rho_2\omega\psi_2$$

the Laplace formula results in the following relation:

$$\rho_1\left(\frac{\omega}{k}\psi_1'(0) + \psi_1(0)V_0'\right) + \rho_2 g\xi_0 - \rho_2\omega\psi_2(0) = -\alpha k^2\xi_0$$

or, taking into account (7.33),

$$\psi_1'(0) \approx \varepsilon^{-1}k\psi_1(0)[1 - (gk + \alpha k^3/\rho_2)/\omega^2]. \tag{7.34}$$

In the zeroth-order approximation in powers of ε this gives the usual dispersion relation for gravitational-capillary waves:

$$\omega = \omega_0(k) = (gk + \alpha k^3/\rho_2)^{1/2}.$$

To obtain the instability let us multiply equation (7.31) by ψ_1^* and subtract the complex-conjugate expression from the result, so that

$$\psi_1^*\frac{d^2\psi_1}{dz^2} - \psi_1\frac{d^2\psi_1^*}{dz^2} = \frac{d}{dz}\left(\psi_1^*\frac{d\psi_1}{dz} - \psi_1\frac{d\psi_1^*}{dz}\right) = 2i\frac{(\mathrm{Im}\,\omega)kV_0''}{|\omega - kV_0|^2}|\psi_1|^2.$$

After the integration of this equality over z from $z = 0$ to $z = +\infty$ we find

$$\left(\psi_1\frac{d\psi_1^*}{dz} - \psi_1^*\frac{d\psi_1}{dz}\right)_{z=0} = 2ik\int_0^\infty \frac{(\mathrm{Im}\,\omega)V_0''}{|\omega - kV_0|^2}|\psi_1|^2\, dz \tag{7.35}$$

(it has been assumed here that ψ_1 tends to zero at infinity). Since for infinitesimal $\mathrm{Im}\,\omega$

$$\lim_{\mathrm{Im}\,\omega\to 0}\frac{\mathrm{Im}\,\omega}{|\omega - kV_0(z)|^2} = \pi\delta[\omega - kV_0(z)]$$

it follows from (7.35) and (7.34) that

$$\operatorname{Im}\left(\psi_1^* \frac{d\psi_1}{dz}\right)_{z=0} = \varepsilon^{-1} k|\psi_1(0)|^2 \operatorname{Im}\left(1 - \frac{\omega_0^2(k)}{\omega^2}\right)$$

$$= -\frac{\pi V_0''(z_r)|\psi_1(z_r)|^2}{|V_0'(z_r)|}$$

where z_r indicates the resonant point with $\omega_0/k = V_0(z_r)$. Writing now $\omega \approx \omega_0(k) + i\gamma$, with $\gamma \ll \omega_0$, we can find after a simple expansion that

$$\gamma \approx -\varepsilon \frac{\pi}{2} \frac{\omega_0(k)}{k} \frac{V_0''(z_r)}{|V_0'(z_r)|} \left|\frac{\psi_1(z_r)}{\psi_1(0)}\right|^2 .$$

Thus a surface wave will be excited if $V_0''(z_r) < 0$. Evidently the air flow is the energy source here, which may be proved by straightforward energy balance considerations that for $V_0''(z_r) < 0$ a surface wave is really created by a decrease of the kinetic energy of the air flow in the vicinity of a resonant elevation.

It was tacitly assumed above that in (7.35) Im $\omega > 0$, so that the δ-function contribution is positive. This is a rather delicate point concerning the singularity in equation (7.31). For a purely inviscid fluid it may be resolved by considering the initial-value problem for ψ with the help of a Laplace transformation, when Im ψ is at once seen to be positive. This results in the well-known *Landau rule* for passing the resonant singularities in the integrals like (7.35) that arose in the eigenmode approach (see problem 85). The same conclusion can be achieved when a singularity is avoided by taking into account a finite viscosity in the fluid equations†.

Problem 61

Calculate the force acting on a sphere of radius a, moving in an inviscid incompressible fluid. Consider the flow over a sphere as the potential.

Let us consider the motion of the sphere to be given, so the coordinate of its centre is $\boldsymbol{R}(t)$. For the potential flow of an incompressible fluid

$$\boldsymbol{v} = \nabla\varphi \qquad \operatorname{div} \boldsymbol{v} = 0 \qquad \text{i.e. } \nabla^2\varphi = 0.$$

This equation for φ must be solved under the boundary condition

$$v_n|_s = \partial\varphi/\partial n|_s = (\dot{\boldsymbol{R}}\cdot\boldsymbol{n})$$

† See, for instance, Lin C C 1955 *The Theory of Hydrodynamic Stability* (Cambridge: Cambridge University Press).

where $v_n|_s$ is the normal component of the fluid velocity at the surface of the sphere, $\dot{\boldsymbol{R}}$ is the sphere velocity and $\boldsymbol{n}$ is a unit vector normal to its surface. It follows from symmetry considerations that the corresponding solution of the Laplace equation for φ has to be of the dipole form, so

$$\varphi(\boldsymbol{r},t) = \frac{\alpha\dot{\boldsymbol{R}}(\boldsymbol{r}-\boldsymbol{R})}{|\boldsymbol{r}-\boldsymbol{R}|^3} = \frac{\alpha\dot{\boldsymbol{R}}\boldsymbol{\rho}}{\rho^3} \qquad \boldsymbol{\rho} = \boldsymbol{r}-\boldsymbol{R}$$

$$\boldsymbol{v} = \nabla\varphi = \frac{\alpha}{\rho^3}[\dot{\boldsymbol{R}} - 3\boldsymbol{n}(\dot{\boldsymbol{R}}\cdot\boldsymbol{n})] \qquad \boldsymbol{n} = \boldsymbol{\rho}/\rho. \tag{7.36}$$

Then the boundary condition $(\boldsymbol{v}\cdot\boldsymbol{n})|_{\rho=a} = (\dot{\boldsymbol{R}}\cdot\boldsymbol{n})$ gives that the coefficient is $\alpha = -a^3/2$. The force of interest is determined by the pressure distribution around a sphere:

$$\mathscr{F}_{\mathrm{f}} = -\int p\,\mathrm{d}\boldsymbol{s} = -\int p\boldsymbol{n}\cdot\mathrm{d}s \tag{7.37}$$

(the vector $\boldsymbol{n}$ is directed outwards from the sphere). For the known velocity field (7.36) the pressure in the fluid may be found from the following equation of motion:

$$\partial\boldsymbol{v}/\partial t + (\boldsymbol{v}\cdot\nabla)\boldsymbol{v} = -\nabla p/\rho_0. \tag{7.38}$$

Since $(\boldsymbol{v}\cdot\nabla)\boldsymbol{v} = \nabla\frac{1}{2}v^2 - \boldsymbol{v}\times\operatorname{curl}\boldsymbol{v}$ and $\operatorname{curl}\boldsymbol{v} = 0$ for a potential flow, it follows from (7.38) that

$$p = p_0 - \rho_0 v^2/2 - \rho_0\,\partial\varphi/\partial t \tag{7.39}$$

where p_0 is the pressure at infinity, where a fluid is at rest. Since

$$\frac{\partial\varphi}{\partial t} = \frac{\alpha\ddot{\boldsymbol{R}}\cdot\boldsymbol{\rho}}{\rho^3} + \frac{2\alpha(\dot{\boldsymbol{R}})^2}{\rho^3}$$

and

$$v^2 = \frac{\alpha^2}{\rho^6}[(\dot{\boldsymbol{R}})^2 + 3(\dot{\boldsymbol{R}}\cdot\boldsymbol{n})^2]$$

it is easy to see that only the term proportional to the acceleration of the sphere, $\ddot{\boldsymbol{R}}$, contributes to the integral in (7.37), so

$$\begin{aligned}\mathscr{F}_{\mathrm{f}} &= -\int p\,\mathrm{d}\boldsymbol{s} = \frac{\alpha\rho_0}{a^2}\int(\boldsymbol{n}\cdot\ddot{\boldsymbol{R}})\,\mathrm{d}\boldsymbol{s} = -\frac{a\rho_0}{2}\int(\boldsymbol{n}\cdot\ddot{\boldsymbol{R}})\boldsymbol{n}\,\mathrm{d}s\\ &= -\tfrac{2}{3}\pi\alpha^3\rho_0\ddot{\boldsymbol{R}}.\end{aligned} \tag{7.40}$$

Thus under a potential flow the force acting on a moving body depends only on its acceleration rather than its velocity. For a moving sphere this force is directed in the opposite sense to the acceleration vector arising from the symmetry, but in the general case it has an oblique direction, so that

$$(\mathscr{F}_{\mathrm{f}})_{\mathrm{i}} = -m_{ik}\ddot{R}_k$$

where m_{ik} is called the induced-mass tensor (see problem 64). As follows from (7.40), for a sphere $m_{ik} = \tilde{m}\delta_{ik}$, with the induced mass

$$\tilde{m} = \tfrac{2}{3}\pi a^3 \rho_0. \tag{7.41}$$

Problem 62

Find the frequency of a mathematical pendulum formed by a small sphere of density ρ_1 immersed in an ideal incompressible fluid with density ρ_0 ($\rho_1 > \rho_0$).

Without a fluid the frequency of the mathematical pendulum is $\omega_0 = (g/l)^{1/2}$, where g is the gravitational acceleration, and l is the length of the pendulum. The fluid has a twofold influence. Firstly, the Archimedes principle results in the effective reduction of gravity by a factor of $(\rho_1 - \rho_0)/\rho_1$. Although the frequency ω_0 is independent of the pendulum mass, in a fluid a pendulum frequency suffers additional reduction because of the fact that the induced mass of the sphere, $\tilde{m}$, contributes effectively only to the inertia of the pendulum, but not to the gravitational force. This reduces the frequency by a factor $[(M + \tilde{m})/M]^{1/2}$, where the sphere mass is $M = \frac{4}{3}\pi a^3 \rho_1$ and the induced mass $\tilde{m}$ is given by (7.41). Finally, we obtain that

$$\omega = (g/l)^{1/2}(1 - \rho_0/\rho_1)^{1/2}(1 + \rho_0/2\rho_1)^{-1/2}.$$

Problem 63

A light cylinder (its density is small in comparison with the density of the surrounding fluid) is immersed coaxially in a vertical pipe of radius R, which is filled with an ideal incompressible fluid. Determine the acceleration as the cylinder floats upwards owing to gravity, if its length L is much larger than R (see figure 7.6).

The cylinder acceleration a will be established at such a value that the Archimedes buoyancy force is balanced by the drag force of the flowing fluid (cylinder weight and inertia are negligible). Thus

$$L\pi r^2 \rho g = \tilde{m}a$$

where ρ is the density of the fluid, and m is the corresponding induced mass. The latter may be obtained in the following way. For an incompressible fluid the motion of the cylinder upwards must be accompanied by a fluid flow in

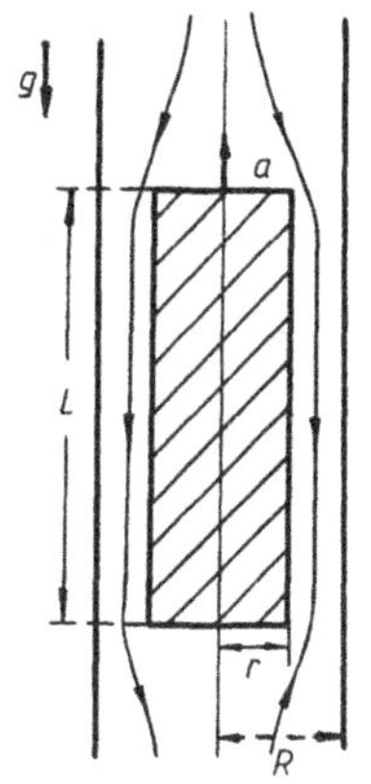

Figure 7.6. Flow around a long cylinder located in a pipe.

the opposite direction. Under the condition $L \gg R$ edge effects are negligible, so this counterflow may be considered as uniform in the hollow cross section of the area $\pi(R^2 - r^2)$ along the whole cylinder. Therefore, the following relation is valid for the cylinder velocity U and the fluid velocity V: $V = Ur^2/(R^2 - r^2)$. So the net momentum of the fluid is

$$P = \rho V\pi(R^2 - r^2)L = \rho\pi r^2 LU$$

and it follows from the momentum conservation law that under the cylinder acceleration $a = \dot{U}$ the corresponding drag force will be

$$\mathscr{F} = -\dot{p} = -\pi r^2 L\rho\dot{U}.$$

Thus the induced mass for this flow is $\tilde{m} = \pi r^2 L\rho$, and the cylinder will float upwards with the acceleration $a = g$.

Problem 64

Derive the induced-mass tensor for the body shown in figure 7.7: a long dumbbell, formed by two spheres of radius a, rigidly connected by a thin bar of a length $l \gg a$.

Let the velocity of the dumbbell be $\boldsymbol{U}$. Then a fluid flow around the sphere 1 results in the additional fluid velocity v_{12} at the location of sphere 2:

$$\boldsymbol{v}_{12} = \frac{a^3}{2l^3}[-\boldsymbol{U} + 3\boldsymbol{n}(\boldsymbol{n}\cdot\boldsymbol{U})] \qquad \boldsymbol{n} = \boldsymbol{l}/l$$

(see problem 61, equation (7.36)). This is equivalent to sphere 2 moving

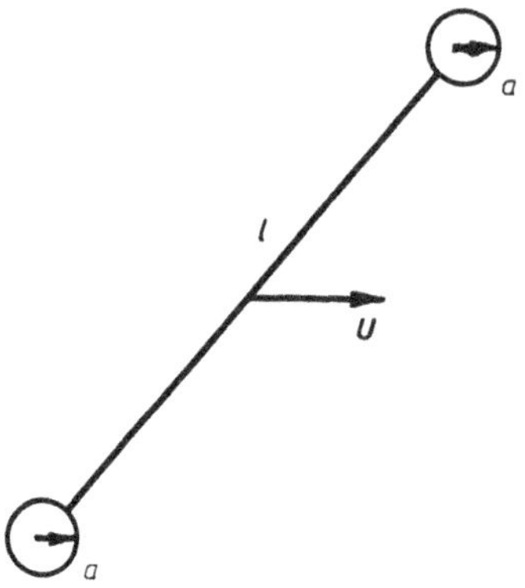

Figure 7.7. On the calculation of the induced-mass tensor for a long dumbbell.

relative to the fluid with the velocity

$$\boldsymbol{U}_2 = \boldsymbol{U} - \boldsymbol{v}_{12} = \boldsymbol{U} + \frac{a^3}{2l^3}[\boldsymbol{U} - 3\boldsymbol{n}(\boldsymbol{n}\cdot\boldsymbol{U})].$$

Since this additional velocity is an even function of $\boldsymbol{n}$, the same result is valid for the first sphere: $\boldsymbol{U}_1 = \boldsymbol{U}_2$. Thus the net force acting on the dumbbell is

$$\begin{aligned}\mathscr{F} &= -\tfrac{2}{3}\pi a^3\rho(\dot{\boldsymbol{U}}_1 + \dot{\boldsymbol{U}}_2)\\ &= -\tfrac{4}{3}\pi a^3\rho[\dot{\boldsymbol{U}}(1 + a^3/2l^3) - 3\boldsymbol{n}(\boldsymbol{n}\cdot\dot{\boldsymbol{U}})a^3/2l^3].\end{aligned} \tag{7.42}$$

By the designation of the induced-mass tensor m_{ik} this force is equal to $\mathscr{F}_i = -m_{ik}\dot{U}_k$. So after rewriting the expression (7.42) in the tensor notation we obtain that

$$m_{ik} = \tfrac{4}{3}\pi a^3\rho\left[\left(1 + \frac{a^3}{2l^3}\right)\delta_{ik} - \frac{3a^3}{2l^3}n_i n_k\right].$$

Chapter 8

The Motion of Viscous Fluids

Recommended textbooks:
Landau L D and Lifshitz E M *Course of Theoretical Physics* vol 6 (Oxford: Pergamon)
Curle N and Davies H J *Modern Fluid Dynamics* (Princeton, NJ: Van Nostrand)

The momentum flux density tensor Π_{ik} takes the following form in a viscous fluid:

$$\Pi_{ik} = p\delta_{ik} + \rho v_i v_k + \pi_{ik} \tag{8.1}$$

where the additional (in comparison with the formula (7.6) for an inviscid fluid) term π_{ik} determines a viscous momentum transport resulting from non-uniformity of the velocity field in a fluid. If the characteristic space scale L of the velocity variation in a fluid or a gas is much larger than the molecular mean free path l, in the first-order approximation in the small parameter l/L the tensor π_{ik} is determined by the first-order derivatives of the velocity vector over the coordinates (spatial derivatives). In an isotropic medium the most general form for π_{ik} is as follows:

$$\pi_{ik} = -\eta\left(\frac{\partial v_i}{\partial x_k} + \frac{\partial v_k}{\partial x_i} - \frac{2}{3}\delta_{ik}\frac{\partial v_l}{\partial x_l}\right) - \zeta\delta_{ik}\frac{\partial v_l}{\partial x_l} \tag{8.2}$$

where η and ζ are the *coefficients of viscosity*, each positive: $\eta \geqslant 0$, $\zeta \geqslant 0$ (the latter is a consequence of the second law of thermodynamics, see problem 65). By substitution of expressions (8.1) and (8.2) into the momentum conservation equation (7.6) and considering the viscosity coefficients η and ζ as constants, we obtain the equation of motion of a viscous fluid:

$$\rho\left(\frac{\partial \boldsymbol{v}}{\partial t} + (\boldsymbol{v}\cdot\nabla)\boldsymbol{v}\right) = -\nabla p + \eta\nabla^2\boldsymbol{v} + (\zeta + \tfrac{1}{3}\eta)\nabla \operatorname{div} \boldsymbol{v}. \tag{8.3}$$

In the particular case of an incompressible fluid, when $\operatorname{div} \boldsymbol{v} = 0$, the *Navier–Stokes equation* follows from (8.3):

$$\partial\boldsymbol{v}/\partial t + (\boldsymbol{v}\cdot\nabla)\boldsymbol{v} = -\nabla p/\rho + \nu\nabla^2\boldsymbol{v} \tag{8.4}$$

where $\nu \equiv \eta/\rho$ is the *kinematic viscosity* of the fluid.

Problem 65

Calculate the energy dissipation rate in an incompressible fluid.

In an incompressible fluid no work is done on the compression (or rarefaction) of the fluid elements. Therefore, the dissipated heat may be derived only from the kinetic energy of the fluid. Thus, let us calculate the rate of change of the fluid kinetic energy inside some fixed volume. Taking into account the equation of motion (8.4) and the condition div $\boldsymbol{v} = 0$ we obtain with the help of the tensor notation:

$$\dot{E}_k = \int\limits_V \rho v_i \frac{\partial v_i}{\partial t}\, \mathrm{d}V = \int\limits_V v_i \left(-\rho v_k \frac{\partial v_i}{\partial x_k} - \frac{\partial p}{\partial x_i} + \eta \frac{\partial^2 v_i}{\partial x_k^2} \right) \mathrm{d}V. \tag{8.5}$$

Since

$$v_i \frac{\partial^2 v_i}{\partial x_k^2} = \frac{\partial}{\partial x_k}\left(v_i \frac{\partial v_i}{\partial x_k} \right) - \left(\frac{\partial v_i}{\partial x_k} \right)^2$$

and

$$\rho v_i \left(v_k \frac{\partial v_i}{\partial x_k} + \frac{1}{\rho}\frac{\partial p}{\partial x_i} \right) = \frac{\partial}{\partial x_k}\left[\rho v_k \left(\frac{v^2}{2} + \frac{p}{\rho} \right) \right]$$

(the relation div $\boldsymbol{v} = \partial v_k / \partial x_k = 0$ is used here) the integral over a volume in (8.5) may be transformed into the form

$$\dot{E}_k = -\int\limits_s \mathrm{d}s_k \left[\rho v_k \left(\frac{v^2}{2} + \frac{p}{\rho} \right) - \eta v_i \frac{\partial v_i}{\partial x_k} \right] - \eta \int \left(\frac{\partial v_i}{\partial x_k} \right)^2 \mathrm{d}V. \tag{8.6}$$

The surface integral in (8.6) describes the change of energy connected with the fluid inflow to a volume (the term $\rho v_k v^2/2$) and with the work produced by surface stresses. So the dissipated power is given by the volume integral in (8.6). Since the latter is always positive, it follows, in particular, that the viscous coefficient $\eta \geqslant 0$. Thus the viscous dissipation rate per unit volume is equal to $Q = \eta (\partial v_i / \partial x_k)^2$.

Problem 66

The plane 'bottom' of an infinitely deep viscous fluid is driven into motion with the velocity $v = v_0 \cos \omega t$. Determine the mean external power needed to maintain these oscillations. The density of the fluid is equal to ρ and its kinetic viscosity is ν.

Let a fluid occupy the domain $z > 0$ with the bottom as the $z = 0$ plane moving along the x-axis. Then only the x-component of the fluid velocity will be present: $v_x(z,t) \equiv v(z,t)$. The following equation for $v(z,t)$ follows from (8.4):

$$\partial v/\partial t = \nu\, \partial^2 v/\partial z^2. \tag{8.7}$$

Since a fluid will oscillate with the same frequency we may find its velocity in the form

$$v(z,t) = \mathrm{Re}[U(z)\mathrm{e}^{-\mathrm{i}\omega t}].$$

Then

$$-\mathrm{i}\omega U = \nu\, \mathrm{d}^2 U/\mathrm{d}z^2 \tag{8.8}$$

and the solution of this equation, which is equal to v_0 at $z = 0$ and dies out at infinity, takes the form:

$$U = v_0 \exp[-(1-\mathrm{i})z/(\omega/2\nu)^{1/2}]$$

therefore

$$v(z,t)$$

$$= v_0 \exp[-z(\omega/2\nu)^{1/2}]\left\{\cos \omega t \cos\left[z\left(\frac{\omega}{2\nu}\right)^{1/2}\right] + \sin \omega t \sin\left[z\left(\frac{\omega}{2\nu}\right)^{1/2}\right]\right\}.$$

The driving external force per unit area of the bottom is equal to the momentum flux:

$$\mathscr{F}_i = (\Pi_{ik} n_k)|_{z=0} = -\Pi_{iz}|_{z=0}$$

so its x-component, which is the only one different from zero, is

$$\mathscr{F}_x = -\rho\nu \left.\frac{\partial v}{\partial z}\right|_{z=0} = v_0 \rho\nu(\omega/2\nu)^{1/2}(\cos \omega t + \sin \omega t).$$

The mean driving power of this force is equal to

$$\langle \mathscr{F}_x v(0,t)\rangle = \rho v_0^2 (\nu\omega/8)^{1/2}$$

(the sign $\langle\ \rangle$ means the time average). Evidently this power is transformed into heat in an oscillating viscous fluid. This may be checked in a straightforward manner by the integration of the dissipation power Q, derived in the previous problem, over the volume per unit area of the bottom.

Problem 67

A plate with the mass per unit area μ is floating in a cuvette on a thin (in comparison with its size) layer of viscous fluid of width equal to h. The bottom of the cuvette oscillates with the frequency

ω and the amplitude A. Find the amplitude of the oscillation of the plate.

Let the oscillations be in the x-direction, and the bottom and the plate correspond to the $z = 0$ and $z = h$ planes. After the notation $v_x(z, t) \equiv v(z)\,e^{i\omega t}$ we find from equation (8.8) that

$$v_x(z,t) = \{v_1 \exp[(1-i)z(\omega/2\nu)^{1/2}] + v_2 \exp[-(1-i)z(\omega/2\nu)^{1/2}]\}\,e^{-i\omega t}$$

where v_1 and v_2 are constants, which have to be determined from the boundary conditions. If the amplitude of the plate oscillation is equal to a, then

$$v_1 + v_2 = -i\omega A$$

$$v_1 \exp[(1-i)h(\omega/2\nu)^{1/2}] + v_2 \exp[-(1-i)h(\omega/2\nu)^{1/2}] = -i\omega a$$

so that

$$v_1 = \frac{-i\omega\{a - A\exp[-(1-i)h(\omega/2\nu)^{1/2}]\}}{\exp[(1-i)h(\omega/2\nu)^{1/2}] - \exp[-(1-i)h(\omega/2\nu)^{1/2}]}$$

$$v_2 = \frac{i\omega\{a - A\exp[(1-i)h(\omega/2\nu)^{1/2}]\}}{\exp[(1-i)h(\omega/2\nu)^{1/2}] - \exp[-(1-i)h(\omega/2\nu)^{1/2}]}.$$

Taking into account the equation of motion for the plate:

$$-\mu\omega^2 a = \mathscr{F}_x = -\rho\nu \left.\frac{\partial v_x}{\partial z}\right|_{z=h}$$

we obtain the plate amplitude a:

$$a = A\Bigg(\frac{\mu(\omega/2\nu)^{1/2}}{(1+i)\rho}\{\exp[(1-i)h(\omega/2\nu)^{1/2}] + \exp[-(1-i)h(\omega/2\nu)^{1/2}]\}$$
$$+ \tfrac{1}{2}\{\exp[(1-i)h(\omega/2\nu)^{1/2}] + \exp[-(1-i)h(\omega/2\nu)^{1/2}]\}\Bigg)^{-1}. \tag{8.9}$$

This rather cumbersome expression may be simplified in some limiting cases connected with the dimensionless parameter $\alpha \equiv \mu/\rho(2\nu/\omega)^{1/2}$. In the case of a 'light' plate, when $\alpha \ll 1$, its amplitude is approximately the same as for a free boundary surface:

$$|a| \approx |A|\left(\frac{2}{\cosh[2h(\omega/2\nu)^{1/2}] + \cos[2h(\omega/2\nu)^{1/2}]}\right)^{1/2}$$

(in this limit the first term in the right-hand side of (8.9) is negligible). For a 'heavy' plate, when $\alpha \gg 1$, the second term in (8.7) becomes essential only for a very thin width of the fluid layer (for $h \ll (2\nu/\omega)^{1/2}$). Under this condition it follows from (8.9) that

$$|a| \approx |A| [\![1 + \alpha^2 \{\cosh[2h(\omega/2\nu)^{1/2}] - \cos[2h(\omega/2\nu)^{1/2}]\}]\!]^{-1/2}.$$

Thus the amplitudes of the plate and the bottom start to differ appreciably for a width of the fluid layer much smaller than the viscous skin depth (at $h \gtrsim \alpha^{-1}(2\nu/\omega)^{1/2}$).

Problem 68

The plane bottom of an infinitely deep viscous incompressible fluid starts to move instantly with the constant velocity v_0. Determine the fluid motion and the resulting frictional force per unit area of the bottom. The fluid density is ρ and the kinematic viscosity is ν.

Using the same coordinate system as in the previous problems we obtain equation (8.7) for the x-component of the fluid velocity vector. Introducing the dimensionless quantity $u = v/v_0$, we arrive at the equation

$$\partial u/\partial t = \nu\, \partial^2 u/\partial z^2 \qquad (8.10)$$

with the following boundary conditions: $u|_{z=0} = 1$, and $u|_{z\to+\infty} = 0$. Since the bottom plane starts to move instantly, there is no parameter in this problem that specifies a temporal or a spatial scale (the only parameter now is the kinematic viscosity ν). So we may conclude that the solution is a *self-similar*. This means that the fluid velocity profiles at different moments can be transformed into each other by the appropriate change of the spatial scale. Since the kinematic viscosity has the dimensional representation $[\nu] = \mathrm{cm^2\,s^{-1}}$, the dimensionless self-similar variable $\xi = z/(\nu t)^{1/2}$ may be introduced (this is the usual diffusion law, when the spatial scale is proportional to $t^{1/2}$). Since the fluid velocity at the bottom is constant ($u|_{z=0} = 1$), the corresponding solutions must be found in the form $u(z,t) = f(\xi)$. Then the following ordinary differential equation for f may be derived from (8.10):

$$-\frac{\xi}{2}\frac{\mathrm{d}f}{\mathrm{d}\xi} = \frac{\mathrm{d}^2 f}{\mathrm{d}\xi^2}.$$

So $\mathrm{d}f/\mathrm{d}\xi = C\exp(-\xi^2/4)$, where C is an unknown constant for the present. Because $f(0) = 1$,

$$f(\xi) = 1 + C\int_0^{\xi} \exp(-y^2/4)\,\mathrm{d}y.$$

Then the constant C is determined from the condition that $f(+\infty) = 0$. Taking into account the result that

$$\int_0^{\infty} \exp(-y^2/4)\,\mathrm{d}y = \sqrt{\pi}$$

we finally obtain

$$v(z,t) = v_0\left(1 - \frac{1}{\sqrt{\pi}} \int_0^{z/(\nu t)^{1/2}} \exp(-y^2/4)\,\mathrm{d}y\right).$$

Thus more and more layers of the fluid become involved in the flow as time progresses (velocity diffuses into the fluid). The net momentum of the fluid increases as $t^{1/2}$. The friction force at the bottom is

$$\mathscr{F} = -\rho\nu \left.\frac{\partial v}{\partial z}\right|_{z=0} = \rho v_0 \left(\frac{\nu}{\pi t}\right)^{1/2}.$$

Problem 69

A vertical pipe of radius R is filled up with a viscous incompressible fluid with the density ρ and viscosity η. A long ($L \gg R$) light cylinder (its density is much less than ρ) is immersed coaxially into this pipe so that only a small gap of width $h \ll R$ is formed between their lateral surfaces. Find the velocity with which the cylinder floats up owing to gravity g.

This velocity will be established at such a value that the Archimedes buoyancy force is balanced by the viscous friction force acting on the lateral surface of the cylinder. Since the gap between the cylinder and the pipe is narrow, a fluid flow there may be considered as a plane flow. Introducing the coordinate x in the radial direction, so that $x = 0$ at the cylinder surface and $x = h$ at the pipe, and neglecting the edge effects, ($L \gg R$), the following equation may be found for a stationary flow of a viscous fluid along the z-axis:

$$-\frac{\mathrm{d}p}{\mathrm{d}z} + \eta\frac{\mathrm{d}^2 v}{\mathrm{d}x^2} = 0. \qquad (8.11)$$

The boundary conditions are: $v(h) = 0$ and $v(0)$ is equal to the cylinder velocity u. The appearance of a pressure gradient $\mathrm{d}p/\mathrm{d}z$ in equation (8.11) becomes evident from the following considerations. If $\mathrm{d}p/\mathrm{d}z$ were equal to zero, a linear velocity profile would be established in the gap (as follows from (8.11)), with the velocity varying from $v = u$ at the cylinder to $v = 0$

at the pipe. So a fluid flow would be directed in the same direction as the cylinder motion. However, this is impossible in a pipe of finite length because of the incompressibility of the fluid. Thus, such a pressure gradient arises in a fluid that is able to push the corresponding amount of a fluid down. This relation takes the following form:

$$\pi R^2 u = -2\pi R \int_0^h v(x)\mathrm{d}x. \tag{8.12}$$

The general solution of equation (8.11) is

$$v(x) = \frac{x^2}{2\eta}\frac{\mathrm{d}p}{\mathrm{d}z} + ax + b.$$

The constants a and b here are determined by the boundary conditions for v, which give

$$b = u \qquad a = -\frac{h}{2\eta}\frac{\mathrm{d}p}{\mathrm{d}z} - \frac{u}{h}.$$

The pressure gradient may now be found from the relation (8.12), so

$$\mathrm{d}p/\mathrm{d}z = 6uR\eta/h^3.$$

The viscous friction force acting on a cylinder is

$$\mathscr{F}_\mathrm{f} = -2\pi RL\eta\left.\frac{\mathrm{d}v}{\mathrm{d}x}\right|_{x=0} = -2\pi RL\eta a \approx 6\pi R^2 Lu\eta/h^2.$$

Equating this to the buoyancy force $\mathscr{F}_\mathrm{A} = \pi R^2 L\rho g$ (where A refers to the Archimedes force) we find that the velocity with which the cylinder floats upwards is equal to $u \approx gh^2/6\nu$. where $\nu = \eta/\rho$ is the kinematic viscosity of the fluid.

Problem 70

Determine the frequency and the damping decrement for a small-amplitude radial pulsation of an ideal gas bubble with radius R and adiabatic specific ratio γ, immersed in a viscous incompressible fluid with pressure p_0, density ρ_0 and viscosity η.

Let the bubble radius r_0 vary by the law: $r_0 = R + a\exp(-\mathrm{i}\omega t)$, where $a \ll R$ is the pulsation amplitude. The induced fluid flow will be in the radial direction with $v_r(r,t) \equiv v(r,t)$. For the incompressible fluid $vr^2 = \text{const}$, so $v(r,t) = A\,\mathrm{e}^{-\mathrm{i}\omega t}/r^2$. At the surface of the bubble the fluid velocity is equal to $\mathrm{d}r_0/\mathrm{d}t$, therefore $A = -\mathrm{i}\omega a R^2$, that is

$$v(r,t) = -\mathrm{i}\omega a R^2\, \mathrm{e}^{-\mathrm{i}\omega t}/r^2.$$

Knowing the fluid velocity the distribution of pressure p_f may be found from the linearized equation of motion

$$\rho_0 \frac{\partial \boldsymbol{v}}{\partial t} = -\boldsymbol{\nabla} p_f + \eta \nabla^2 \boldsymbol{v}. \tag{8.13}$$

Since $\nabla^2 \boldsymbol{v} = \boldsymbol{\nabla}\,\mathrm{div}\,\boldsymbol{v} - \mathrm{curl\,curl}\,\boldsymbol{v}$, in this case $\nabla^2 \boldsymbol{v} = 0$ ($\mathrm{div}\,\boldsymbol{v} = 0$, $\mathrm{curl}\,\boldsymbol{v} = 0$), and it follows from (8.13) that

$$\frac{\partial p_f}{\partial r} = -\rho_0 \frac{\partial v}{\partial t} = \mathrm{i}\omega\rho_0 v = \rho_0 \omega^2 a R^2\, \mathrm{e}^{-\mathrm{i}\omega t}/r^2.$$

By integration of this equation over r and taking into account that the unperturbed pressure at infinity is equal to p_0, we find

$$p_f(r,t) = p_0 - \rho_0 \omega^2 a R^2\, \mathrm{e}^{-\mathrm{i}\omega t}/r.$$

The gas density inside a bubble is much less than ρ_0, so the bubble compressions and rarefactions may be considered as quasistatic, that is the gas density and pressure remain approximately uniform inside the whole bubble. Therefore, from the adiabaticity condition we have:

$$p_g V^\gamma = p_g r_0^{3\gamma} = \mathrm{const} = p_0 R^{3\gamma}.$$

So the gas pressure varies as follows:

$$p_g = p_0 - 3\gamma p_0 a\, \mathrm{e}^{-\mathrm{i}\omega t}/R.$$

Knowing the pressure in a gas and in a fluid we can write down the boundary condition at their interface at $r = R$ (in a linear approximation the bubble radius variation may be neglected here). The latter consists in the continuity of the momentum flux vector $\sigma_{ik} n_k$, where σ_{ik} is the tension tensor and $\boldsymbol{n}$ is the unit vector normal to the boundary. In a gas $\sigma_{ik}^{(g)} = -p_g \delta_{ik}$, while in a viscous fluid

$$\sigma_{ik}^{(f)} = -p_f \delta_{ik} + \eta(\partial v_i/\partial x_k + \partial v_k/\partial x_i).$$

In our case this boundary condition results in the equality

$$\sigma_{rr}^{(g)} = \sigma_{rr}^{(f)}|_{r=R}$$

which gives the following equation for ω:

$$\omega^2 + \frac{4\mathrm{i}\eta}{\rho_0 R^2} - \frac{3\gamma p_0}{\rho_0 R^2} = 0$$

or

$$\omega = -\frac{2\mathrm{i}\eta}{\rho_0 R^2} \pm \left(\frac{3\gamma p_0}{\rho_0 R^2} - \frac{4\eta^2}{\rho_0 R^4}\right)^{1/2}.$$

When the viscosity is small (when $\eta < (\frac{3}{4}\gamma p_0 \rho_0 R^2)^{1/2}$) a bubble oscillates with a frequency of

$$\Omega = \left(\frac{3\gamma p_0}{\rho_0 R^2} - \frac{4\eta^2}{\rho_0 R^4} \right)^{1/2}$$

and a damping decrement of $\Gamma = 2\eta/\rho_0 R^2$. In an inviscid fluid the bubble eigenfrequency is equal to $\Omega = (3\gamma p_0/\rho_0 R^2)^{1/2}$. In the case of a rather large viscosity the bubble radial displacement is damped aperiodically.

Chapter 9

Convection and Turbulence

Recommended textbooks:
Landau L D and Lifshitz E M *Fluid Mechanics* (Oxford: Pergamon)
Turner J S *Buoyancy Effects in Fluids* (Cambridge: Cambridge University Press)
Landahl M T and Mollo-Christensen E *Turbulence and Random Processes in Fluid Mechanics* (Cambridge: Cambridge University Press)

Problem 71

An ideal gas with a molecular weight μ and an adiabatic specific ratio γ is in equilibrium in the gravitation field $\boldsymbol{g}$. Determine the criterion for convective instability in such a gas.

Let us write the equations of motion, continuity and adiabaticity for a gas:

$$\rho\, \mathrm{d}\boldsymbol{v}/\mathrm{d}t = -\nabla p + \rho \boldsymbol{g} \qquad \partial\rho/\partial t + \operatorname{div}(\rho\boldsymbol{v}) = 0$$
$$\frac{\mathrm{d}}{\mathrm{d}t}(p\rho^{-\gamma}) = 0. \tag{9.1}$$

For the state of static equilibrium ($\boldsymbol{v} = 0$) it follows from (9.1) that

$$-\nabla p + \rho\boldsymbol{g} = 0$$

that is

$$\mathrm{d}p/\mathrm{d}z = -\rho g$$

so the pressure, the density (and the temperature of the gas) are functions only of the z-coordinate; $p(z)$, $\rho(z)$ (and $T(z)$). For the analysis of the stability of this equilibrium let us consider small deviations from this state, described by the perturbation of the pressure δp, the perturbation of the density $\delta\rho$, and the velocity of a gas $\boldsymbol{v}$. After the linearization procedure that takes into account only the quantities up to the first order in the perturbations we obtain from (9.1) that

$$\rho\,\partial \boldsymbol{v}/\partial t = -\boldsymbol{\nabla}\delta p + \boldsymbol{g}\delta\rho \qquad \partial\delta\rho/\partial t + \boldsymbol{v}\cdot\boldsymbol{\nabla}\rho + \rho\boldsymbol{\nabla}\cdot\boldsymbol{v} = 0$$
$$\frac{\partial}{\partial t}[\delta p\rho^{-\gamma} - \gamma p\delta\rho\rho^{-(\gamma+1)}] + \boldsymbol{v}\cdot\boldsymbol{\nabla}(p\rho^{-\gamma}) = 0. \tag{9.2}$$

Now introducing the vector $\boldsymbol{\xi}(\boldsymbol{r}, t)$, the displacement of a gas element from the equilibrium position, and taking into account that $\boldsymbol{v} = \partial\boldsymbol{\xi}/\partial t$, we obtain from the linearized continuity and adiabaticity equations (9.2) that

$$\delta\rho = -\rho\,\mathrm{div}\,\boldsymbol{\xi} - \xi_z\,\mathrm{d}\rho/\mathrm{d}z$$

and the perturbation of the pressure

$$\begin{aligned}\delta p &= \gamma p\delta\rho/\rho - \xi_z\left(\frac{\mathrm{d}p}{\mathrm{d}z} - \frac{\gamma p}{\rho}\frac{\mathrm{d}\rho}{\mathrm{d}z}\right)\\ &= \frac{\gamma p}{\rho}\left(-\rho\,\mathrm{div}\,\boldsymbol{\xi} - \xi_z\frac{\mathrm{d}\rho}{\mathrm{d}z}\right) - \xi_z\left(\mathrm{d}p/\mathrm{d}z - \frac{\gamma p}{\rho}\frac{\mathrm{d}\rho}{\mathrm{d}z}\right)\\ &= -\gamma p\,\mathrm{div}\,\boldsymbol{\xi} - \xi_z\,\mathrm{d}p/\mathrm{d}z = -\gamma p\,\mathrm{div}\,\boldsymbol{\xi} + \rho g\xi_z.\end{aligned} \tag{9.3}$$

With the help of these expressions the equation of motion in (9.2) takes the form:

$$\rho\ddot{\boldsymbol{\xi}} = \boldsymbol{g}(-\rho\,\mathrm{div}\,\boldsymbol{\xi} - \xi_z\,\mathrm{d}\rho/\mathrm{d}z) - \boldsymbol{\nabla}(-\gamma p\,\mathrm{div}\,\boldsymbol{\xi} + \rho g\xi_z) \equiv \mathscr{F}(\boldsymbol{\xi}) \tag{9.4}$$

where $\mathscr{F}(\boldsymbol{\xi})$ is a linear functional of $\boldsymbol{\xi}$ and means the force arising from this displacement. The following energy conservation law results from equation (9.4):

$$W_\mathrm{k} + W_\mathrm{p} = \mathrm{const}$$

where

$$W_\mathrm{k} = \frac{1}{2}\int \rho(\dot{\boldsymbol{\xi}})^2\,\mathrm{d}V$$

is the kinetic energy of the gas and

$$W_\mathrm{p} = -\frac{1}{2}\int \mathrm{d}V\mathscr{F}(\boldsymbol{\xi})\cdot\boldsymbol{\xi}$$

is the potential one. Therefore, it now follows from the general principles of mechanics that the equilibrium state of a gas will be stable if and only if the potential energy W_p, considered as a functional of the infinite number of degrees of freedom $\boldsymbol{\xi}(\boldsymbol{r})$, has a minimum at $\boldsymbol{\xi}(\boldsymbol{r}) \equiv 0$. In other words, in the stable situation $W_p \geqslant 0$ for an arbitrary distribution of $\boldsymbol{\xi}(\boldsymbol{r})$ consistent with the prescribed boundary condition (see below). In our case

$$W_{\mathrm{p}} = -\frac{1}{2}\int \mathscr{F}\cdot\boldsymbol{\xi}\,\mathrm{d}V = \frac{1}{2}\int \mathrm{d}V\left[(\boldsymbol{\xi}\cdot\boldsymbol{g})\left(\rho\,\mathrm{div}\,\boldsymbol{\xi} + \xi_z\frac{\mathrm{d}\rho}{\mathrm{d}z}\right)\right.$$
$$\left. + \boldsymbol{\xi}\cdot\nabla(-\gamma p\,\mathrm{div}\,\boldsymbol{\xi} + \rho g\xi_z)\right]. \tag{9.5}$$

The second term in (9.5) may be transformed as follows:

$$\int \mathrm{d}V\boldsymbol{\xi}\cdot\nabla(-\gamma p\,\mathrm{div}\,\boldsymbol{\xi} + \rho g\xi_z) = \int \mathrm{d}V\,\mathrm{div}[\boldsymbol{\xi}(-\gamma p\,\mathrm{div}\,\boldsymbol{\xi} + \rho g\xi_z)]$$
$$- \int \mathrm{div}\,\boldsymbol{\xi}(-\gamma p\,\mathrm{div}\,\boldsymbol{\xi} + \rho g\xi_z)\mathrm{d}V$$
$$= \int (-\gamma p\,\mathrm{div}\,\boldsymbol{\xi} + \rho g\xi_z)\boldsymbol{\xi}\,\mathrm{d}\boldsymbol{S}$$
$$- \int (-\gamma p\,\mathrm{div}\,\boldsymbol{\xi} + \rho g\xi_z)\,\mathrm{div}\,\boldsymbol{\xi}\,\mathrm{d}V$$

Since at the rigid boundary $\boldsymbol{\xi}\cdot\mathrm{d}\boldsymbol{S} = 0$ (the normal component of the displacement is absent), and at infinity p and ρ are equal to zero, the surface integral in this expression may be omitted, so that

$$W_{\mathrm{p}} = \frac{1}{2}\int \mathrm{d}V\left[-g\xi_z\left(\rho\,\mathrm{div}\,\boldsymbol{\xi} + \xi_z\frac{\mathrm{d}\rho}{\mathrm{d}z}\right) + \gamma p(\mathrm{div}\,\boldsymbol{\xi})^2 - \rho g\xi_z\,\mathrm{div}\,\boldsymbol{\xi}\right]$$
$$= \frac{1}{2}\int \mathrm{d}V\left(\gamma p(\mathrm{div}\,\boldsymbol{\xi})^2 - 2\rho g\xi_z\,\mathrm{div}\,\boldsymbol{\xi} - g(\xi_z)^2\frac{\mathrm{d}\rho}{\mathrm{d}z}\right). \tag{9.6}$$

Let us now search for the minimum of W_{p} by considering it as a functional of ξ_z and div $\boldsymbol{\xi}$. From the extremum condition on div $\boldsymbol{\xi}$ we obtain that

$$2\gamma p\,\mathrm{div}\,\boldsymbol{\xi} - 2\rho g\xi_z = 0$$

that is

$$\mathrm{div}\,\boldsymbol{\xi} = \rho g\xi_z/\gamma p. \tag{9.7}$$

So it follows from (9.6) and (9.7) that for this case

$$W_{\mathrm{p}} = \frac{1}{2}\int \mathrm{d}V\,\xi_z^2\left(-g\frac{\mathrm{d}\rho}{\mathrm{d}z} - \frac{\rho^2 g^2}{\gamma p}\right). \tag{9.8}$$

It is now seen that a necessary and sufficient condition for the stability is the positivity of the expression in brackets in (9.8). Indeed, if it is negative at some point, then it is possible to make W_{p} also negative by choosing ξ_z to be localized at this point. So we may conclude that the criterion for the convective stability of an ideal gas has a local character and is as follows:

$$d\rho/dz < -\rho^2 g/\gamma p.$$

Since for a gas $p = R\rho T/\mu$

$$\frac{d\rho}{dz} = \frac{\mu}{RT}\frac{dp}{dz} - \frac{\mu p}{RT^2}\frac{dT}{dz}$$

and $dp/dz = -\rho g$, this criterion may be written in the form:

$$-dT/dz < \mu g(\gamma - 1)/\gamma R.$$

It is seen that stable equilibrium may also be possible for a gas heated from below, but only if the gradient of the temperature is not too large.

It is worth considering now the physical nature of the most 'dangerous' perturbations found above. It follows from equations (9.7) and (9.3) that $\delta p = 0$ for this case. So this means that they represent the quasistatic interchange of the two neighbouring elements of a gas with slightly different vertical positions.

Problem 72

Find the criterion for the convective instability in an incompressible fluid.

The term 'incompressible fluid' means here that the density of the fluid, which is, generally, a function of the pressure and the temperature, $\rho(p, T)$, has a very weak dependence on the pressure, so we may consider approximately that $\rho \equiv \rho(T)$. If the temperature of a fluid varies in the vicinity of some value T_0, this dependence may be written in the form: $\rho(T) \approx \rho_0[1 - \alpha(T - T_0)]$, where α is the thermal-expansion coefficient of the fluid (we shall assume that $\alpha > 0$). Since the resulting density variation is also small, it may be neglected in the continuity equation, which allows us to put $\operatorname{div} \boldsymbol{v} = 0$. In the equation of motion for the fluid

$$\frac{d\boldsymbol{v}}{dt} = -\frac{1}{\rho}\nabla p + \boldsymbol{g} + \nu\nabla^2\boldsymbol{v} \tag{9.9}$$

we shall separate from the total pressure p the contribution determined by the hydrostatic equilibrium of the fluid: $p = \rho_0 \boldsymbol{g}\cdot\boldsymbol{r} + p'$, so the correction p' connected with the thermal expansion and the motion of the fluid, may be considered as small. Then

$$-\frac{1}{\rho}\nabla p = \frac{-\nabla(\rho_0 \boldsymbol{g}\cdot\boldsymbol{r} + p')}{\rho_0[1 - \alpha(T - T_0)]} \approx -\boldsymbol{g}[1 + \alpha(T - T_0)] - \frac{\nabla p'}{\rho_0}$$

and equation (9.9) takes the form

$$d\boldsymbol{v}/dt = -\nabla p'/\rho_0 - \alpha\boldsymbol{g}(T - T_0) + \nu\nabla^2\boldsymbol{v}. \tag{9.10}$$

To close the system we need to derive the equation for a temperature variation in a fluid (the thermal conduction equation). The thermal flux in a fluid $\boldsymbol{q}$ consists both in diffusive (thermal conductivity) and convective (fluid flow) contributions and thus has the form

$$\boldsymbol{q} = \rho_0 C T \boldsymbol{v} - \kappa \nabla T$$

where κ is the thermal conductivity of the fluid, and C is the specific heat capacity. Then the heat balance equation

$$\partial(\rho_0 C T)/\partial t + \operatorname{div} \boldsymbol{q} = 0$$

results in

$$\rho_0 C \frac{\partial T}{\partial t} = -\operatorname{div}(\rho_0 C T \boldsymbol{v} - \kappa \nabla T) = -\rho_0 C \boldsymbol{v} \cdot \nabla T + \kappa \nabla^2 T$$

(the coefficients C and κ are considered as constant because of the weak variation of the temperature, and the equation $\operatorname{div} \boldsymbol{v} = 0$ is taken into account). Therefore

$$\frac{\partial T}{\partial t} + \boldsymbol{v} \cdot \nabla T = \chi \nabla^2 T \qquad (9.11)$$

($\chi \equiv \kappa/\rho_0 C$ is the fluid thermometric conductivity). Now equations (9.10), (9.11) and the condition $\operatorname{div} \boldsymbol{v} = 0$ form a complete system describing convection in a fluid (commonly called the Boussinesq equations).

Let the fluid be immersed between two horizontal planes $z = 0$ and $z = L$ with the fixed temperatures T_0 and T_1, respectively (figure 9.1). In the static equilibrium the temperature profile $T(z)$ satisfies, according to (9.11), the

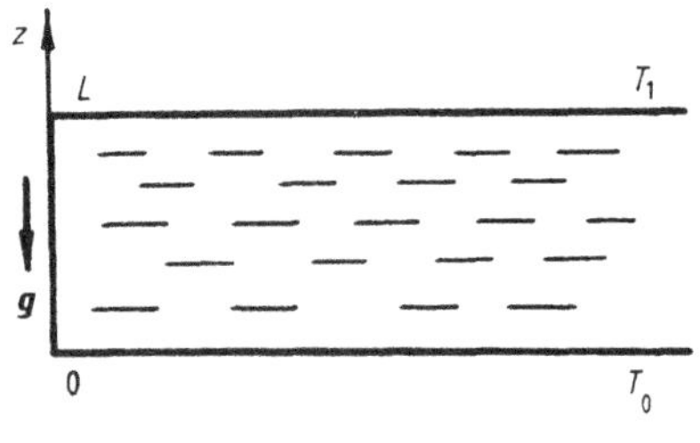

Figure 9.1. Convection in a fluid.

equation $d^2T/dz^2 = 0$, so that

$$T(z) = T_0 + (T_1 - T_0)z/L.$$

Therefore the equilibrium condition from (9.10) determines the following pressure distribution:

$$p' = \rho_0 \alpha g (T_1 - T_0) z^2/2L.$$

Now we shall consider the small deviations from this equilibrium state, described by the temperature perturbation δT, the pressure perturbation δp and the velocity of the fluid $\boldsymbol{v}$. In a linear approximation it follows from equations (9.10) and (9.11) that

$$\begin{aligned} \operatorname{div} \boldsymbol{v} &= 0 \\ \partial \boldsymbol{v}/\partial t &= -\rho_0^{-1} \nabla \delta p - \alpha \boldsymbol{g} \delta T + \nu \nabla^2 \boldsymbol{v} \\ \partial \delta T/\partial t + v_z (T_1 - T_0)/L &= \chi \nabla^2 \delta T. \end{aligned} \tag{9.12}$$

It is useful to rewrite this system in terms of dimensionless variables measuring lengths in terms of L, time intervals of L^2/ν, velocities of ν/L, pressures of $\rho_0 \nu^2/L^2$, and temperature perturbations of $T_1 - T_0$. Then equations (9.12) take the form

$$\begin{aligned} \operatorname{div} \boldsymbol{v} &= 0 \\ \partial \boldsymbol{v}/\partial t &= -\nabla \delta p + \nabla^2 \boldsymbol{v} - G \delta T \boldsymbol{e}_z \\ \partial \delta T/\partial t + v_z &= P^{-1} \nabla^2 \delta T \end{aligned} \tag{9.13}$$

(to simplify these expressions we use the same notation for the dimensionless variables), where

$$G \equiv \alpha g L^3 (T_0 - T_1)/\nu^2 \qquad P = \nu/\chi$$

are the *Grashof* and the *Prandtl* numbers, respectively. Thus the problem of convection in an incompressible fluid is characterized by these two dimensionless numbers. This means that this problem possesses *similarity*, that is for different values of the fluid density ρ_0, the thermal-expansion coefficient α and the viscosity of the fluid ν, etc, all the properties of the process will remain the same, provided the numbers G and P remain unchanged (and $\boldsymbol{v}$, δp, δT, etc are measured in the units mentioned above).

Since the equilibrium state is stationary and uniform in the transverse coordinates $\boldsymbol{r}_\perp = (x, y)$, the dependence of the perturbed quantities on $\boldsymbol{r}_\perp$ and t may be taken in the form proportional to $\exp(\lambda t + i\boldsymbol{k}_\perp \cdot \boldsymbol{r}_\perp)$. So the necessary and sufficient condition for the stability is that Re λ is negative for all values of $\boldsymbol{k}_\perp$. At a given value of $\boldsymbol{k}_\perp$ it is always possible to choose the coordinate axis in the x–y plane in such a way that $\boldsymbol{k}_\perp = (k, 0)$. In this case only the v_x and v_z components of the velocity vector are present. Since $\operatorname{div} \boldsymbol{v} = ikv_x + dv_z/dz = 0$, we have $v_x = (i/k)\, dv_z/dz$. Taking now the curl (its y-component) of the equation of motion in (9.13) to eliminate the pressure perturbation (curl $\nabla \delta p = 0$), we find

$$\lambda\left(\frac{dv_x}{dz} - ikv_z\right) = \left(-k^2 + \frac{d^2}{dz^2}\right)\left(\frac{dv_x}{dz} - ikv_z\right) + ikG\delta T$$

or, by expressing v_x over v_z,

$$\lambda\left(\frac{d^2v_z}{dz^2}-k^2v_z\right)=\left(-k^2+\frac{d^2}{dz^2}\right)\left(\frac{d^2v_z}{dz^2}-k^2v_z\right)+k^2G\delta T. \quad (9.14)$$

The thermal conductivity equation in (9.13) takes the form

$$\lambda\delta T+v_z=P^{-1}\left(-k^2+\frac{d^2}{dz^2}\right)\delta T. \quad (9.15)$$

These two equations, (9.14) and (9.15), for v_z and δT have to be solved with the boundary conditions at the bounding planes $z=0$ and $z=1$. Since the latter are kept at constant temperatures, $\delta T|_{z=0,1}=0$. As to the boundary condition for the velocity we shall assume the plane surfaces to be ideally slippery, which means the lack of friction force at $z=0, 1$. Therefore

$$\left.\frac{dv_x}{dz}\right|_{z=0,1}=\left.\frac{i}{k}\frac{d^2v_z}{dz^2}\right|_{z=0,1}=0$$

(in addition, $v_z|_{z=0,1}=0$). It is easy to check that the appropriate solutions of equation (9.14) and (9.15) are as follows:

$$\delta T=A\sin n\pi z \qquad v_2=B\sin n\pi z \qquad n=1,2,\ldots.$$

The compatibility condition determines the eigenvalues λ:

$$\lambda=-\frac{(k^2+n^2\pi^2)}{2}(1+P^{-1})\pm\left[\frac{(k^2+n^2\pi^2)^2}{4}(1+P^{-1})^2+\left(G\frac{k^2}{(k^2+n^2\pi^2)}-\frac{(k^2+n^2\pi^2)^2}{P}\right)\right]^{1/2}.$$

It is seen now that the perturbations grow ($\mathrm{Re}\,\lambda>0$) if the number $R\equiv GP>(k^2+n^2\pi^2)^3/k^2\equiv\mathscr{F}(k,n)$, where $R\equiv\alpha gL^3(T_0-T_1)/\gamma\chi$ is the *Rayleigh number*. Perturbations with different values of n and k have different instability thresholds. At a given value of n the 'most unstable' perturbation is the one with $k=k_*=n\pi/\sqrt{2}$ (it corresponds to the minimum of the function $\mathscr{F}(k,n)$), which increases at $R>R_*(n)=\mathscr{F}(k_*,n)=\frac{27}{4}n^4\pi^4$. Therefore, while the Rayleigh number increases (for instance, with an increase of the temperature difference (T_0-T_1) between the bounding planes) the perturbations with $n=1$ and $k_\perp=\pi/L\sqrt{2}$ are the first to become unstable, and the critical Rayleigh number for convective instability is $R_{cr}=\frac{27}{4}\pi^4\approx 658$.

Thus the convective instability threshold in an incompressible fluid is determined by the one dimensionless parameter (the Rayleigh number). This result can be interpreted in a rather simple physical way, which becomes clear from the following energy balance considerations. The energy source of convective instability is the decrease of a fluid potential energy in a gravity field by the displacement of a hot (and, hence, more light) fluid upwards while a cold (and more heavy) fluid flows downwards. At a given distance L between the bounding planes the characteristic size of the interchanging fluid volumes is of the same order, so the corresponding potential energy gain may be estimated as $\Delta W_p \sim \Delta m g L$, where $\Delta m \sim (\Delta \rho) L^3 \sim \alpha \rho_0 (T_0 - T_1) L^3$ is the mass difference of these volumes. However, the displacement motion in a viscous fluid will be followed by the dissipation of energy and heat production, so some work, ΔW_d, must be performed for the interchange of fluid volumes. Therefore, the convective instability will arise under the potential energy gain ΔW_p exceeding the dissipated energy ΔW_d. The last may be estimated as follows. In an incompressible fluid the dissipation rate per unit volume is equal to $Q = \eta (\partial v_i / \partial x_k)^2$ (see problem 65), so

$$\Delta W_d \sim Q L^3 \Delta t \sim \eta \frac{v^2}{L^2} L^3 \frac{L}{v} \sim \rho_0 \nu v L^2$$

where $\Delta t \sim L/v$ is the characteristic duration of the interchange process. It seems at the first glance that the dissipated energy $\Delta W_d \sim \rho_0 \nu v L^2$, which is proportional to the velocity v, may be made arbitrarily small by reducing v. However, this is not the case owing to the thermal conduction in a fluid. In the estimate given above for ΔW_p it has been tacitly assumed that while displacing the fluid volume, the temperature (and hence, the density) of the fluid remain unchanged and equal to their initial values. However, if the displacement is performed too slowly the temperature of the fluid volumes will be equalized by the thermal conductivity and the potential energy gain will tend to zero. Hence this interchange has to be carried out rather quickly, so that thermal conductivity will not have enough time to become apparent. By comparing the convective $(\boldsymbol{v} \cdot \nabla T)$ and diffusion $(\chi \nabla^2 T)$ terms in equation (9.11) we may conclude that this requires the velocity to be rather large: $v \gtrsim \chi / L$. In this case the dissipated energy $\Delta W_d \sim \rho_0 \nu \chi L$, so convection will arise if $\Delta W_p \sim \alpha \rho_0 (T_0 - T_1) g L^4 > \rho_0 \gamma \chi L$, or $\alpha g L^3 (T_0 - T_1) / \nu \chi > 1$. Thus the Rayleigh number alone determines the convective instability threshold in an incompressible fluid.

Problem 73

Determine the possible types of ordered structures (convective cells) arising for convection close to the marginal state (when the Rayleigh number is just above its critical value: $R - R_{cr} \ll R_{cr}$).

When the Rayleigh number becomes just above the critical value, the most unstable perturbations (that with $n = 1$ and $k_\perp = k_0 = \pi / L\sqrt{2}$ for the model considered above) are excited first. Hence in this case a fluid motion may be represented as a superposition of these modes, so, for instance,

$$v_z(x, y, z) = \sin\left(\frac{\pi z}{L}\right) \operatorname{Re} \int A(k_x, k_y) \exp[\mathrm{i}(k_x x + k_y y)]$$
$$\times\ \delta(k_x^2 + k_y^2 - k_0^2)\, \mathrm{d}k_x\, \mathrm{d}k_y. \tag{9.16}$$

For an arbitrary set of the amplitudes $A(k_x, k_y)$ the resulting motion will be rather irregular. However it is well known experimentally that marginal convection possesses the steady ordered structure periodic in the x–y plane. The whole space between the bounding planes is divided into adjacent identical cells, in each of which the fluid moves in closed paths without passing from one cell to another. The outlines of these cells on the planes form some kind of two-dimensional lattice. What is the possible symmetry of this lattice? It is limited to those regular polygons that may pack a plane closely. Therefore the cross section of a convective cell may be only a rectangle (which degenerates into a square or a band in extreme cases), a regular triangle, and a regular hexagon. Let us consider now what set of amplitudes $A(k_x, k_y)$ in (9.16) relate to each of these structures and determine the periodicity of the corresponding lattices.

The rectangular structure is formed by the superposition of two basic vectors in the k_x–k_y plane with equal amplitudes. Indeed, consider two vectors ($\boldsymbol{k}_1$ and $\boldsymbol{k}_2$) of the same length k_0 and with an angular separation of α (figure 9.2). In the coordinate system shown on this figure

$$\boldsymbol{k}_1 = [k_0 \cos(\alpha/2), k_0 \sin(\alpha/2)] \qquad \boldsymbol{k}_2 = [k_0 \cos(\alpha/2), -k_0 \sin(\alpha/2)].$$

Then for $A_1 = A_2 = v_0/2$ the vertical component of the fluid velocity is

$$v_z = \frac{v_0}{2} \sin\left(\frac{\pi z}{L}\right) [\cos(k_{1x}x + k_{1y}y) + \cos(k_{2x}x + k_{2y}y)]$$
$$= v_0 \sin\left(\frac{\pi z}{L}\right) \cos\left(k_0 x \cos\frac{\alpha}{2}\right) \cos\left(k_0 y \sin\frac{\alpha}{2}\right). \tag{9.17}$$

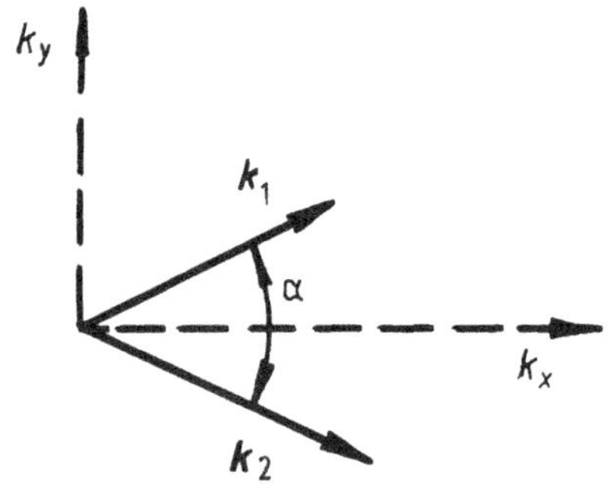

Figure 9.2. Basic wave vectors for a rectangular convective cell.

It follows from the solution given in the previous problem that for a given $\boldsymbol{k}_\perp$ the horizontal velocity of the fluid is parallel to $\boldsymbol{k}_\perp$. Therefore in this case

$$\boldsymbol{v}_\perp = \boldsymbol{v}_{\perp 1} + \boldsymbol{v}_{\perp 2} = -\frac{\pi}{k_0 L}\frac{v_0}{2}\cos\left(\frac{\pi z}{L}\right)$$
$$\times\left(\frac{\boldsymbol{k}_1}{k_0}\sin(k_{1x}x + k_{1y}y) + \frac{\boldsymbol{k}_2}{k_0}\sin(k_{2x}x + k_{2y}y)\right)$$

or, by components,

$$\begin{aligned} v_x &= -v_0\frac{\pi\cos\alpha/2}{k_0 L}\cos\left(\frac{\pi z}{L}\right)\sin\left(k_0 x\cos\frac{\alpha}{2}\right)\cos\left(k_0 y\sin\frac{\alpha}{2}\right) \\ v_y &= -v_0\frac{\pi\sin\alpha/2}{k_0 L}\cos\left(\frac{\pi z}{L}\right)\cos\left(k_0 x\cos\frac{\alpha}{2}\right)\sin\left(k_0 y\sin\frac{\alpha}{2}\right). \end{aligned} \tag{9.18}$$

The resulting horizontal streamlines and lines with $v_z(x, y) = 0$ are shown in figure 9.3. The dashed lines correspond to $v_z = 0$, while the bold arrowed ones represent the particular (separatrix) streamlines in the x–y plane that correspond to the boundaries of the convective cells at $z = L$. We assume $v_0 > 0$, and the + and − signs in the figure correspond to the domains where the fluid flows up and down. Thus, the convective cell is a rectangular prism with a height equal to L, and with sides $\Delta x = \pi/k_0 \cos\alpha/2, \Delta y = \pi/k_0 \sin\alpha/2$ (at $\alpha \to 0$ this cell degenerates into the band along the y-axis with the width $\Delta x = \pi/k_0$). Each cell is divided into four parts (see the figure): in two of them a fluid flows up, and in the other two down.

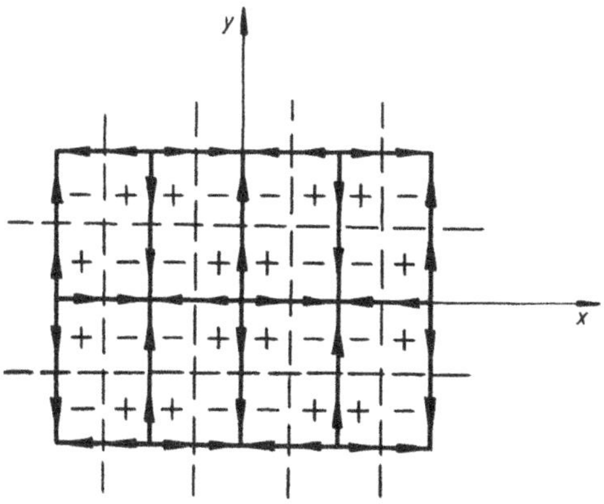

Figure 9.3. Structure of a rectangular convective lattice.

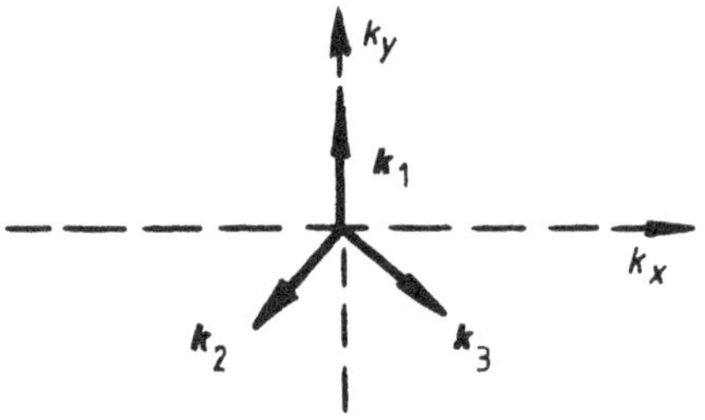

Figure 9.4. Basic wave vectors for the triangular convective cell.

Let us go to a triangular lattice. In this case the solution is a superposition of the three basic vectors $\boldsymbol{k}_1$, $\boldsymbol{k}_2$ and $\boldsymbol{k}_3$ (figure 9.4): $\boldsymbol{k}_1 = \boldsymbol{k}_0(0, 1)$:

$$\boldsymbol{k}_2 = k_0(-\sqrt{3}/2, -\tfrac{1}{2}) \qquad \boldsymbol{k}_3 = k_0(\sqrt{3}/2, -\tfrac{1}{2})$$

with the amplitudes $A_1 = A_2 = A_3 = -iv_0$. Then it follows from (9.16) that

$$v_z = v_0 \sin\left(\frac{\pi z}{L}\right)\left\{\sin k_0 y + \sin\left[k_0\left(\frac{\sqrt{3}}{2}x - \frac{y}{2}\right)\right] - \sin\left[k_0\left(\frac{\sqrt{3}}{2}x + \frac{y}{2}\right)\right]\right\}$$

$$v_x = v_0 \frac{\sqrt{3}}{2}\frac{\pi}{k_0 L}\cos\left(\frac{\pi z}{L}\right)\left[\cos k_0\left(\frac{\sqrt{3}}{2}x - \frac{y}{2}\right) - \cos k_0\left(\frac{\sqrt{3}}{2}x + \frac{y}{2}\right)\right]$$

$$v_y = \frac{v_0}{2}\frac{\pi}{k_0 L}\cos\left(\frac{\pi z}{L}\right)\left[2\cos k_0 y - \cos k_0\left(\frac{\sqrt{3}}{2}x + \frac{y}{2}\right) - \cos k_0\left(\frac{\sqrt{3}}{2}x - \frac{y}{2}\right)\right].$$

The resulting pattern in the x–y plane is shown in figure 9.5. The lines with $v_z(x, y) = 0$ (the dashed lines in the figure) are the straight lines dividing the x–y plane on the regular triangles with a side equal to $4\pi/k_0\sqrt{3}$. In each of them the fluid flows up (+) or down (−). The separatrix horizontal streamlines corresponding to the boundaries of the convective cells at $z = L$ are shown by the bold lines with the arrows along the fluid horizontal velocity. It is seen that each cell represents a regular triangular prism with a side equal to $4\pi/3k_0$. The cells are divided into two equal parts where the fluid flows up and down.

The hexagonal lattice (called Bernard cells) is formed by the same basic vectors $\boldsymbol{k}_1$, $\boldsymbol{k}_2$ and $\boldsymbol{k}_3$ as the triangular structure, but with different amplitudes: $A_1 = A_2 = A_3 = v_0$. Thus

$$v_z = v_0 \sin\frac{\pi z}{L}\left[\cos k_0 y + \cos k_0\left(\frac{\sqrt{3}}{2}x + \frac{y}{2}\right) + \cos k_0\left(\frac{\sqrt{3}}{2}x - \frac{y}{2}\right)\right]$$

$$v_x = -\frac{\sqrt{3}}{2}v_0\frac{\pi}{k_0 L}\cos\left(\frac{\pi z}{L}\right)\left[\sin k_0\left(\frac{\sqrt{3}}{2}x + \frac{y}{2}\right) + \sin k_0\left(\frac{\sqrt{3}}{2}x - \frac{y}{2}\right)\right]$$

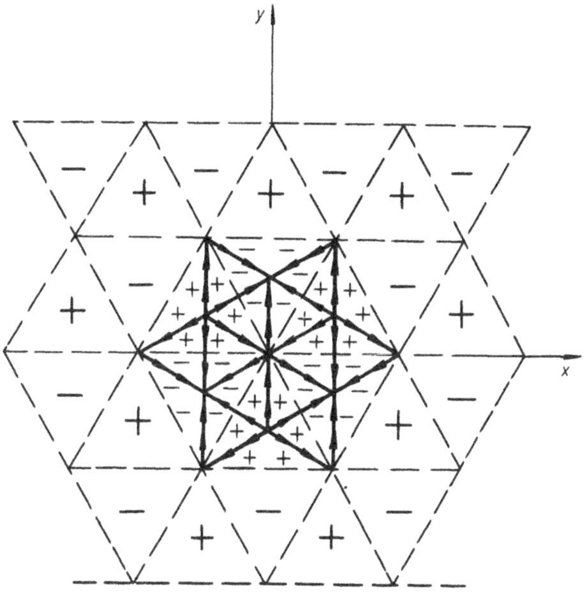

Figure 9.5. Streamlines for a triangular convective lattice.

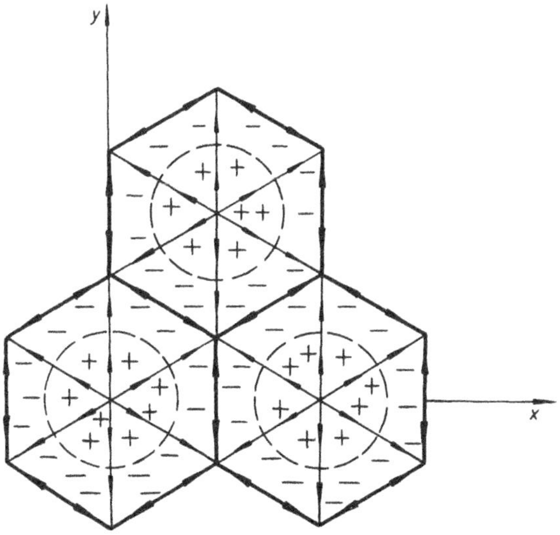

Figure 9.6. Structure of an hexagonal convective lattice.

$$v_y = -\frac{v_0}{2}\frac{\pi}{k_0 L}\cos\frac{\pi z}{L}\left[2\sin k_0 y + \sin k_0\left(\frac{\sqrt{3}}{2}x + \frac{y}{2}\right) - \sin k_0\left(\frac{\sqrt{3}}{2}x - \frac{y}{2}\right)\right].$$

The corresponding flow pattern is shown in figure 9.6. The separatrix horizontal streamlines are the same as in the previous triangular case but now they are not the cell boundaries because of the difference in the vertical velocity, v_z. The lines with $v_z(x, y) = 0$ represent the set of isolated closed curves (dashed lines in the figure), very similar to the circles. Inside them a fluid flows up (we put $v_0 > 0$), and outside it flows down. Therefore, the unit structural element of a lattice in the x–y plane is now a regular hexagon with a side equal to $4\pi/3k_0$. So each convective cell represents a hexagonal prism with the fluid flowing up at the central part and down near the boundaries.

Problem 74

> Derive the relation between the convective heat flux (directed from the lower bounding plane to the upper one) and the fluid kinetic energy for the case of square convective cells in the regime of marginal convection ($R - R_{cr} \ll R_{cr}$).

The convective heat flus is $\boldsymbol{q}^{(v)} = \rho_0 c T \boldsymbol{v}$, and its vertical component, which is of interest to us, is $q_z^{(v)} = \rho_0 c T \boldsymbol{v}_z$. Since the net vertical fluid flux inside each convective cell is equal to zero, the net convective heat flux inside a cell will be proportional to the square of the convective motion amplitude:

$$Q = \int q_z^{(v)}\, \mathrm{d}x\, \mathrm{d}y = \rho_0 c S \langle \delta T v_z \rangle$$

where δT is the temperature perturbation under convection, S is the cross section area of the cell, and the brackets mean the averaging over the cell cross sectional area. The value of δT can be expressed over v_z with the help of the heat conductivity equation in (9.12), where the time derivative term $\partial \delta T/\partial t$ may be neglected for the marginal regime:

$$v_z \frac{(T_1 - T_0)}{L} = \chi \Delta T = -\chi\left(k_0^2 + \frac{\pi^2}{L^2}\right)\delta T$$

that is

$$\delta T = [2L(T_0 - T_1)/3\pi^2\chi]v_z$$

(we have put $k_0^2 = \pi^2/2L^2$ here). As follows from the previous problem, for the square cell the sides are equal to $\Delta x = \Delta y = 2L$, and the vertical velocity, according to (9.17),

$$v_z = v_0 \sin\frac{\pi z}{L} \cos\frac{\pi x}{2L} \cos\frac{\pi y}{2L}\,.$$

Thus

$$\langle \delta T v_z \rangle = \frac{(T_0 - T_1)Lv_0^2}{6\pi^2 \chi} \sin^2 \frac{\pi z}{L}$$

and the net convective heat flux

$$Q = \frac{2\rho_0 c v_0^2 L^3 (T_0 - T_1)}{3\pi^2 \chi} \sin^2 \frac{\pi z}{L}.$$

The incoming velocity amplitude v_0 has to be derived from the non-linear equations for steady convection. It follows from rather general considerations† that in the marginal regime, when $R - R_{cr} \ll R_{cr}$, its value is proportional to $(R - R_{cr})^{1/2}$, so that the net convective heat flux is $Q \propto (R - R_{cr})$ (this is the so-called 'soft' regime of a steady convection).

Problem 75

Determine the order of magnitude v_τ of a velocity variation over a time interval τ of a portion of fluid as it moves about in a fully developed turbulent flow. It is considered that τ is short compared with the time $T_L \sim L/u$ characterizing the flow as a whole.

Under the fully developed turbulence a fluid chaotic motion may be represented as a superposition of turbulent eddies of different sizes λ. In the *inertial interval*: $\lambda_0 \lesssim \lambda \lesssim L$, where L is the space scale of the mean flow, and λ_0 is the size where the viscosity of the fluid becomes important, the order of magnitude of the velocity v_λ of the eddy with given size λ is determined by the *Kolmogorov–Obukhov law*: $v_\lambda \sim u(\lambda/L)^{1/3}$ (u is the mean flow velocity). Therefore, the portion of fluid is involved simultaneously in the motions of eddies of different sizes. Let us consider the effect of some eddy with given λ on the interested velocity variation. This effect depends appreciably on the relation between the considered time interval τ and the characteristic period corresponding to this eddy $T_\lambda \sim \lambda/v_\lambda \sim T_L(\lambda/L)^{2/3}$. For $\tau \gg T_\lambda$ (which holds true for a rather short-scale eddy) the velocity variation $v_\tau(\lambda)$ will be of the order of the eddy velocity: $v_\tau(\lambda) \sim v_\lambda \sim u(\lambda/L)^{1/3}$. On the contrary, in a case when $\tau \ll T_\lambda$ (for a sufficiently long λ) the velocity variation of the portion of the fluid will form only the small part of v_λ:

$$v_\tau(\lambda) \sim v_\lambda \frac{\tau}{T_\lambda} \sim \frac{u^2}{L} \tau \left(\frac{\lambda}{L}\right)^{-1/3}.$$

† See, for instance, Landau L D and Lifshitz E M *Course of Theoretical Physics* vol 6 (Oxford: Pergamon) §26.

Thus it becomes clear from these estimates that the main contribution to v_τ will be carried by the eddies with a characteristic period of the same order as τ: $T_{\lambda *} \sim \tau$, that is $\lambda_* \sim L(\tau/T_L)^{3/2}$. Then

$$v_\tau \sim v_{\lambda *} \sim u(\tau/T_L)^{1/2}.$$

This result will still be valid only for τ not too short, so that the corresponding λ_* lies inside the inertial interval: $\lambda_* \sim L(\tau/T_L)^{3/2} \gg \lambda_0$. Otherwise, if $\tau < T_L(\lambda_0/L)^{2/3}$, the motion of the portion of fluid will be regular during this time interval, so its velocity variation will be proportional to τ.

Problem 76

Derive the variation in time of the distance between two adjacent fluid particles under a fully developed turbulent flow.

Let the distance between two fluid particles at some instant be l, with $\lambda_0 \ll l \ll L$. Their subsequent separation will be determined by the difference of their velocities. To estimate the latter we consider the contribution of turbulent eddies of different scales λ. If $\lambda \lesssim l$, the corresponding velocity difference v_l will be of the same order of magnitude as v_λ: $v_l \sim v_\lambda \sim u(\lambda/L)^{1/3}$. On the contrary, for a rather large eddy, when $\lambda \gtrsim l$ we obtain that $v_l \sim v_\lambda l/\lambda \sim u(l/L)(\lambda/L)^{-2/3}$. Thus the main role is played by the eddies with $\lambda \sim l$, so the velocity difference is $v_l \sim u(l/L)^{1/3}$. It is worth noting here that the contribution of the mean flow to this velocity will be of the order $ul/L \ll v_l$, and therefore may be neglected. So the distance between two fluid particles will vary with the law: $dl/dt \sim v_l \sim u(l/L)^{1/3}$. This relation shows that the divergence between adjacent fluid particles in a turbulent flow has a self-accelerating nature, so after some time the distance between the two fixed fluid particles becomes independent of their initial separation and grows according to the law: $l(t) \sim u^{3/2}t^{3/2}/L^{1/2}$.

Problem 77

In a pipe of radius a and length L a steady flow of incompressible fluid of density ρ and viscosity η is maintained by the pressure difference Δp between the ends of the pipe. Determine the fluid discharge (the mass of fluid passing per unit time through any cross section of the pipe) for a small value of Δp, when the flow is laminar, and under a large pressure difference (in a fully developed turbulent regime).

For a steady laminar flow the following equation for the fluid velocity $v_z(r) \equiv v(r)$ results from the Navier–Stokes equation (8.4):

$$\frac{1}{r}\frac{\mathrm{d}}{\mathrm{d}r}\left(r\frac{\mathrm{d}v}{\mathrm{d}r}\right) = -\frac{\Delta p}{\eta L}.$$

Its regular solution, equal to zero at the pipe wall ($r = a$), is

$$v(r) = \Delta p(a^2 - r^2)/4\eta L.$$

Thus the velocity profile has a parabolic form (figure 9.7(a)), and the fluid discharge is

$$Q = 2\pi\rho\int_0^a vr\,\mathrm{d}r = \pi(\Delta p)a^4/8\nu L \tag{9.19}$$

(the Poiseuille formula).

Let us now consider the turbulent flow. It may be described in terms of the z-component of momentum flux, σ, on the wall of the pipe, which is equal to the frictional force on unit area of the pipe surface. It follows from the force balance:

$$\pi a^2\Delta p = \sigma 2\pi aL \qquad \text{that} \qquad \sigma = (\Delta p)a/2L.$$

Under given σ the mean flow velocity variation near the wall has the logarithmic form†:

$$v(y) = \frac{1}{\kappa}\left(\frac{\sigma}{\rho}\right)^{1/2}\ln\left[\frac{y}{\nu}\left(\frac{\sigma}{\rho}\right)^{1/2}\right] \tag{9.20}$$

where y is the distance from the wall and κ is a numerical constant ($\kappa \approx 0.4$). Such a logarithmic profile is valid for $y \ll a$, until the momentum flux is approximately constant (this formula fails only at the wall, since $v(0) = 0$). However, for the turbulent flow when the Reynolds number $R \equiv av/\nu$ is large, this formula gives the correct result (with logarithmic accuracy) inside the whole cross section. The reason is that even for $y \ll a$ the argument under the logarithm in (9.20) becomes large compared with unity, so that expression (9.20) possesses a very weak variation while y increases. Indeed, at $y \sim a$ this argument is $(a^3\Delta p/\rho L\nu^2)^{1/2}$. From the other side, for the laminar flow, the characteristic velocity of the fluid is by the order of magnitude equal to $v \sim a^2\Delta p/\rho\nu L$, so the Reynolds number is $R = av/\nu \sim a^3\Delta p/\rho\nu^2 L$. Thus it is seen that in the main part of cross section the argument under the logarithm is of the order of $R^{1/2}$, and because the transition to a turbulent flow occurs at $R \gg 1$, this argument is also large. Summing up we may conclude that for the turbulent flow in a pipe the mean velocity of a fluid is approximately

† See, for instance, Landau L D and Lifshitz E M *Course of Theoretical Physics* vol 6 (Oxford: Pergamon) §42.

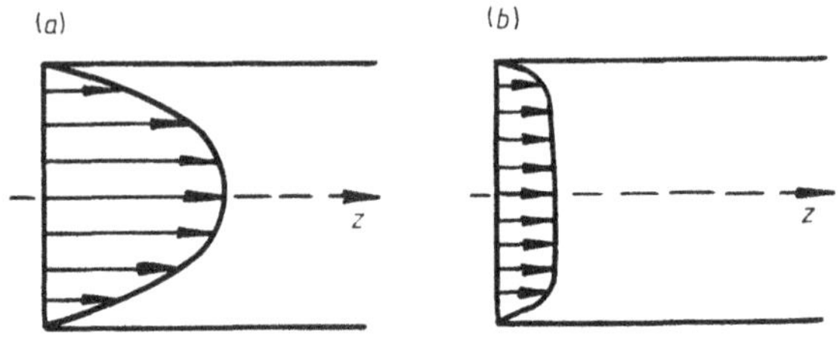

Figure 9.7. Fluid velocity profiles for (*a*) laminar and (*b*) turbulent flow in a pipe.

uniform in the main part of cross section area (figure 9.7(*b*)) and equal to

$$v \sim \frac{1}{\kappa}(a\Delta p/2L\rho)^{1/2} \ln\left(\frac{a}{\nu}(a\Delta p/2L\rho)^{1/2}\right).$$

The fluid discharge is

$$Q \approx \pi a^2 \rho v \approx \frac{\pi a^2 \rho}{\kappa}(a\Delta p/2L\rho)^{1/2} \ln\left(\frac{a}{\nu}(a\Delta p/2L\rho)^{1/2}\right).$$

Let us note that while in the laminar regime the fluid discharge is proportional to the first power of the pressure difference, for the turbulent flow $Q \propto (\Delta p)^{1/2}$, so it becomes appreciably smaller than for the laminar flow under the same Δp.

Problem 78

Estimate the spatial scale of the motions, where a viscous energy dissipation occurs, for the main part of the cross sectional area of a pipe with the turbulent flow considered in the previous problem.

For the main part of the pipe the difference between the mean velocities is of the order of $(\sigma/\rho)^{1/2} \sim (a\Delta p/\rho L)^{1/2}$. Therefore the largest eddies have a spatial scale $L \sim a$ and the velocity amplitude $u \sim (a\Delta p/\rho L)^{1/2}$. Then, according to the Kolmogorov–Obukhov law, in the inertial interval, the following relationship holds:

$$v_\lambda \sim u(\lambda/a)^{1/3} \sim (a\Delta p/\rho L)^{1/2}(\lambda/a)^{1/3}.$$

The lower boundary of the inertial interval, λ_0, where the viscosity of the fluid comes into play, may be determined by the relation that the corresponding effective Reynolds number for these eddies,

$$R(\lambda_0) \sim \lambda_0 v(\lambda_0)/\nu \sim (a\Delta p/\rho L)^{1/2}(\lambda_0/a)^{1/3}\lambda_0/\nu$$

becomes of the order of unity. So it follows that

$$\lambda_0 \sim a(\nu^2 \rho L/a^3 \Delta p)^{3/8}.$$

Chapter 10

Acoustic Waves

Recommended textbooks:
Landau L D and Lifshitz E M *Fluid Mechanics* (Oxford: Pergamon)
Whitham G B *Linear and Nonlinear Waves* (New York: Wiley)

In a linear (acoustic) approximation the propagation of the sound waves in a medium may be described by the following wave equation for the velocity potential φ ($\boldsymbol{v} = -\nabla\varphi$):

$$\partial^2\varphi/\partial t^2 - c_0^2\nabla^2\varphi = 0 \tag{10.1}$$

where the velocity of sound c_0 is determined by the adiabatic compressibility of the medium:

$$c_0^2 = (\partial p/\partial\rho)_s.$$

For a plane travelling wave propagating in the direction specified by the unit vector $\boldsymbol{n}$, the potential has the following dependence on the coordinates and time:

$$\varphi(\boldsymbol{r}, t) = f(\boldsymbol{r}\cdot\boldsymbol{n} - c_0 t) \tag{10.2}$$

In this case the density and the pressure perturbations in a medium ($\delta\rho$ and δp) relate in a unique fashion with the velocity $\boldsymbol{v} = \boldsymbol{n}v$:

$$\delta\rho = \rho v/c_0 \qquad \delta p = \rho v c_0 \qquad v = -f'. \tag{10.3}$$

The energy density E, the energy flux density $\boldsymbol{q}$ and the tensor of the momentum flux density Π_{ik} in such a wave are equal to

$$E = \rho v^2 \qquad \boldsymbol{q} = c_0 E\boldsymbol{n} \qquad \Pi_{ik} = En_i n_k. \tag{10.4}$$

Problem 79

Determine the motion arising in an ideal gas from the initial perturbation described by the velocity distribution $v_x(x) \equiv u(x)$ and pressure $p(x) = p_0 + a(x)$. Consider that $a(x)$ and $u(x)$ tend to zero at infinity and are sufficiently small that the acoustic approximation is valid.

It is evident that the subsequent motion in a gas will be one dimensional, so the linearized continuity equation and the equation of motion may be written in the forms

$$\frac{\partial \delta\rho}{\partial t} + \rho\frac{\partial v}{\partial x} = 0 \qquad \rho\frac{\partial v}{\partial t} = -\frac{\partial \delta p}{\partial x} \tag{10.5}$$

where $\delta\rho$ and δp, the density and the pressure perturbations, are related to each other by the adiabaticity condition:

$$\delta p = (\partial p/\partial \rho)_s \delta\rho = c_0^2 \delta\rho.$$

Then the following wave equation follows from (10.5) for the velocity of a gas:

$$\partial^2 v/\partial t^2 - c_0^2\,\partial^2 v/\partial x^2 = 0.$$

Its general solution may be easily obtained by the introduction of the new variables $\xi = x - c_0 t$ and $\eta = x + c_0 t$. In these variables the wave equation for v takes the form: $\partial^2 v/\partial\xi\,\partial\eta = 0$, so its general solution is

$$v(\xi, \eta) = f(\xi) + g(\eta)$$

where f and g are arbitrary functions. Reverting to the variables (x, t), we have

$$v(x, t) = f(x - c_0 t) + g(x + c_0 t).$$

The functions f and g have to be determined from the initial conditions at $t = 0$, when $v(x, 0) = u(x)$ and

$$\left.\frac{\partial v}{\partial t}\right|_{t=0} = -\frac{1}{\rho}\frac{\partial \delta p(x, 0)}{\partial x} = -\frac{1}{\rho}\frac{\mathrm{d}a}{\mathrm{d}x}.$$

Then it follows that

$$f(x) + g(x) = u(x) \qquad c_0\left(\frac{\mathrm{d}g}{\mathrm{d}x} - \frac{\mathrm{d}f}{\mathrm{d}x}\right) = -\frac{1}{\rho}\frac{\mathrm{d}a}{\mathrm{d}x}.$$

Since all the perturbations disappear at infinity, we finally find that

$$v(x, t) = \tfrac{1}{2}[u(x - c_0 t) + u(x + c_0 t)] + \frac{1}{2c_0\rho}[a(x - c_0 t) - a(x + c_0 t)]. \tag{10.6}$$

Problem 80

The plane compression wave with pulse length l and a linear pressure profile varying from p_m to zero (figure 10.1 (a)) propagates in a fluid. When this wave reaches the free boundary, which is

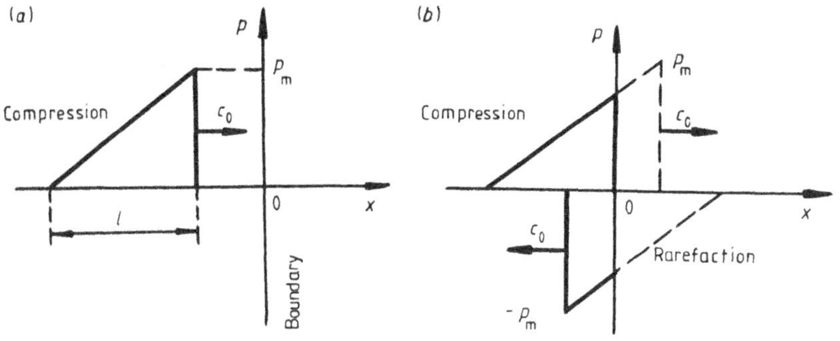

Figure 10.1. Superposition of compressive and rarefactive pulses.

parallel to the wave front, the rarefaction wave with a negative pressure starts to propagate from the boundary inside the fluid to maintain the boundary condition $p = 0$ at the free surface (figure 10.1(b)). If a negative pressure in a fluid exceeds some critical value p_{cr}, the fluid divides at this place. How many fluid layers will be formed because of the division and what will be their width, if $p_m = np_{cr}$ (n an integer)? Consider the acoustic approximation to be valid.

The kinematics of the waves becomes clear from figure 10.1(b). Let the compression wave reach the boundary at the instant $t = 0$. Since both waves propagate at the sound velocity c_0 (but in opposite directions), at the instant $t = l/2nc_0$ at the position $x = -l/2n$ the net pressure will be equal to $-p_{cr} = -p_m/n$, so the splitting of a fluid layer of width $\delta = l/2n$ will occur. After that we return to a situation similar to the initial one, with the only difference that the maximum compression pressure is now equal to $p'_m = (n-1)p_{cr}$. Therefore the total number of divided fluid layers will be equal to n with the width $\delta = l/2n$ each.

Problem 81

The monochromatic sound wave, propagating in a fluid with density ρ_1 and sound velocity c_1, is reflected normally from the plane boundary with another fluid of density ρ_2 and sound velocity c_2. Derive the reflection coefficient for this wave.

Because of reflection three monochromatic sound waves with frequency ω equal to that of the incident wave will propagate in the fluids (figure 10.2): the incident and reflected waves in the fluid 1, and the refracted wave in the fluid 2. Denoting their amplitudes as A_1, A_2 and A_3 respectively, we obtain

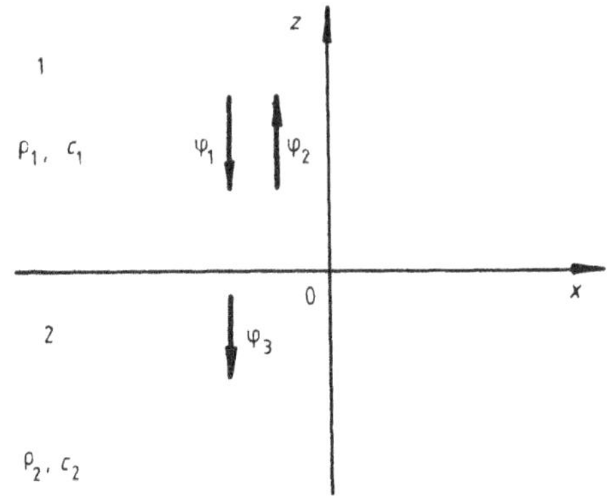

Figure 10.2. Reflection of a sound wave.

for the velocity potentials:

$$\varphi_1 = A_1 \exp[i\omega(-z/c_1 - t)] \qquad \varphi_2 = A_2 \exp[i\omega(z/c_1 - t)]$$
$$\varphi_3 = A_3 \exp[i\omega(-z/c_2 - t)].$$

We use here the relation $\boldsymbol{k} = \omega \boldsymbol{n}/c$ for sound waves ($\boldsymbol{k}$ is the wave vector and $\boldsymbol{n} = \boldsymbol{k}/k$). At the boundary surface (at $z = 0$) the continuity of the pressure and of the normal velocity component must be satisfied, which relates the amplitudes A_2 and A_3 with A_1. It follows from the linearized equation of motion, $\rho\, \partial \boldsymbol{v}/\partial t = -\nabla \delta p$, that the pressure perturbation δp in a sound wave is equal to $\delta p = \rho\, \partial\varphi/\partial t = -i\omega\rho\varphi$. Thus the boundary conditions take the form:

$$i\omega\rho_1(\varphi_1 + \varphi_2)|_{z=0} = i\omega\rho_2\varphi_3|_{z=0}$$
$$(\partial\varphi_1/\partial z + \partial\varphi_2/\partial z)_{z=0} = \partial\varphi_3/\partial z|_{z=0}$$

so

$$A_2 = A_1 \frac{c_2\rho_2 - c_1\rho_1}{c_1\rho_1 + c_2\rho_2} \qquad A_3 = A_1 \frac{2c_2\rho_1}{c_1\rho_1 + c_2\rho_2}.$$

The reflection coefficient R is equal to the ratio of the energy flux densities in the reflected and incident waves. Therefore, it follows from (10.4) that

$$R = \frac{q_2}{q_1} = \left(\frac{A_2}{A_1}\right)^2 = \left(\frac{c_2\rho_2 - c_1\rho_1}{c_1\rho_1 + c_2\rho_2}\right)^2.$$

Problem 82

An ideal gas with temperature T, molecular weight μ and adiabatic specific ratio γ is located in a vertical pipe of length L under the gravitational field $\boldsymbol{g}$. Both ends of the pipe are closed. Determine the eigenfrequencies of the longitudinal (vertical) oscillations of the gas.

In the steady equilibrium state the gas pressure varies with height by the law $\mathrm{d}p/\mathrm{d}z = -\rho g$. Together with the equation of state for an ideal gas: $p = R\rho T/\mu$ this results in the well known barometric formula for the density and pressure distribution:

$$\rho = \rho_0\,\mathrm{e}^{-z/h} \qquad p = RT\rho_0\mu^{-1}\,\mathrm{e}^{-z/h} \qquad h = RT/\mu g.$$

The vertical oscillations in a gas result in the density perturbation $\delta\rho(z,t)$, the pressure perturbation $\delta p(z,t)$ and the gas velocity $v_z(z,t) \equiv v(z,t)$. Consider all the perturbed quantities proportional to $\mathrm{e}^{-\mathrm{i}\omega t}$, we obtain from the linearized continuity equation, the equation of motion and the entropy conservation the following relations:

$$-\mathrm{i}\omega\delta\rho + \frac{\partial}{\partial z}(v\rho_0\,\mathrm{e}^{-z/h}) = 0 \tag{10.7}$$

$$-\mathrm{i}\omega\rho_0\,\mathrm{e}^{-z/h}v = -\partial\delta p/\partial z - g\delta\rho \tag{10.8}$$

$$\frac{\mathrm{d}S}{\mathrm{d}t} = \frac{\partial\delta S}{\partial t} + v\frac{\partial S_0}{\partial z} = 0 \tag{10.9a}$$

or, since $S \propto p\rho^{\gamma}$ (where S is entropy)

$$-\mathrm{i}\omega\left(\frac{\delta p}{\rho^{\gamma}} - \frac{\gamma p}{\rho^{\gamma+1}}\delta\rho\right) + v\frac{\partial}{\partial z}(p\rho^{-\gamma}) = 0. \tag{10.9b}$$

Expressing now $\delta\rho$ and δp over v with the help of equations (10.7) and (10.9) and by substitution of the result into (10.8), the following equation may be obtained for the quantity $u(z) \equiv v(z)\,\mathrm{e}^{-z/h}$:

$$-\omega^2 u = c_0^2\frac{\mathrm{d}^2 u}{\mathrm{d}z^2} + \gamma g\frac{\mathrm{d}u}{\mathrm{d}z}$$

where $c_0 = (\gamma RT/\mu)^{1/2}$ is the sound velocity in this gas. Since the coefficients in this equation are independent of z, its solution may be found in the form $u(z) \propto \mathrm{e}^{\mathrm{i}kz}$. Then it follows at once that

$$k = \mathrm{i}\gamma g/2c_0^2 \pm (\omega^2/c_0^2 - \gamma^2 g^2/4c_0^4)^{1/2}.$$

So the general solution is

$$u(z) = A\exp(-\gamma gz/2c_0^2)\exp[\mathrm{i}z(\omega^2/c_0^2 - \gamma^2 g^2/4c_0^4)^{1/2}]$$
$$+ B\exp(-\gamma gz/2c_0^2)\exp[-\mathrm{i}z(\omega^2/c_0^2 - \gamma^2 g^2/4c_0^4)^{1/2}].$$

From the boundary condition $u(0) = 0$ we find that $B = -A$, and the second boundary condition, $u(L) = 0$, determines the eigenfrequencies ω:

$$(\omega^2/c_0^2 - \gamma^2 g^2/4c_0^4)^{1/2} = n\pi L \qquad n = 1, 2, \ldots$$

so

$$\omega_n = (\gamma^2 g^2/4c_0^2 + n^2\pi^2 c_0^2/L^2)^{1/2}.$$

It is interesting to note that the velocity of the gas for the eigenmode has the form

$$v(z) = v_0\, e^{z/2h} \sin n\pi z/L$$

so for $L \gg h$ the amplitude of the gas oscillation grows sharply in the upper part of the gas column, where the density is small.

Problem 83

Estimate the decrement of the 'radiation' damping of the small-amplitude radial oscillations of a gas bubble immersed in an inviscid fluid (see problem 70). The radius of the bubble is R, the adiabatic specific ratio of the gas is γ, and the fluid density and pressure are ρ and p, respectively.

The radiation damping of the bubble radial oscillations is determined by the energy losses arising from the emission of sound waves (the latter occurs when the finite compressibility of the fluid is taken into account). To proceed with our estimate we shall suggest (and this will be confirmed by the final result) that this damping is weak, so that the corresponding decrement Γ is small compared with the oscillation frequency ω. The latter has been obtained in problem 70: $\omega = (3\gamma p/\rho R^2)^{1/2}$. In an incompressible fluid the variation of the bubble radius: $r_0(t) = R + a\, e^{-i\omega t}$, results in the fluid flow

$$v_r(r, t) \equiv v(z, t) = -i\omega a R^2\, e^{-i\omega t}/r^2$$

which adjusts to the variation of $r_0(t)$ in the whole space (since in an incompressible fluid the sound velocity $c_0 \to \infty$). If the finite value of c_0 is taken into account, such an adjustment has time to establish only at distances $r \lesssim r_* \sim c_0/\omega$ (in a 'quasistationary' zone). At $r > r_*$ (in a 'wave' zone) the fluid motion represents spherically expanding sound waves. Therefore the amplitude of the sound wave at the boundary of these two regions ($r \sim r_*$) may be estimated as

$$\tilde{v}_* \sim v(r_*) \sim \omega a R^2/r_*^2.$$

Knowing the amplitude $\tilde{v}_*$ it is easy to calculate the net energy flux S carried away by the sound waves. Since, according to (10.4), the energy flux density

$q \sim c_0 \rho \tilde{v}_*^2$, we have

$$S \sim q r_*^2 \sim \rho c_0 \omega^2 a^2 R^4 / r_*^2 \sim \rho \omega^4 a^2 R^4 / c_0.$$

The damping rate is $\Gamma \sim S/\mathscr{E}$, where $\mathscr{E}$ is the energy of the oscillations. In an order of magnitude calculation $\mathscr{E}$ may be estimated as the kinetic energy of the fluid involved in the motion caused by the bubble pulsation. So

$$\mathscr{E} \sim \int \rho v^2 \, \mathrm{d}V \sim \int_R^\infty \rho \omega^2 a^2 R^4 \frac{\mathrm{d}r}{r^2} \sim \rho \omega a^2 R^3$$

and therefore $\Gamma \sim S/\mathscr{E} \sim \omega^2 R / c_0$ and the ratio $\Gamma/\omega \sim \omega R / c_0 \sim (3\gamma p / \rho c_0^2)^{1/2}$. Since the sound velocity in the fluid, c_0, is large compared with $(p/\rho)^{1/2}$, we find that $\Gamma/\omega \ll 1$, thus such a radiation damping is indeed weak.

Problem 84

Derive the damping decrement for the sound wave in a gas owing to thermal conduction. The molecular weight of the gas is μ, the adiabatic specific ratio $-\gamma$ and the thermal conductivity $-\kappa$.

Let us consider the plane monochromatic wave propagating along the z-axis. Then the perturbations of the density $\delta\rho$, of the pressure δp and the gas flow velocity $v_z \equiv v$ are proportional to $\exp(ikz - i\omega t)$, so the linearized equations of continuity, of the motion and of the entropy production may be written as follows:

$$-i\omega\delta\rho + ik\rho v = 0 \qquad i\omega\rho v = ik\delta p \qquad -i\omega\rho T\delta S = \delta Q. \tag{10.10}$$

Here δS is the variation of the entropy per unit mass, and δQ is the heat dissipation rate per unit volume resulting from thermal conduction in the gas. Since the entropy per unit mass for the ideal gas with the adiabatic specific ratio γ is equal to†

$$S = \{R/[\mu(\gamma - 1)]\} \ln(p/\rho\gamma)$$

and the dissipation power is

$$\delta Q = \operatorname{div}(\kappa \nabla T) = \kappa \Delta T = -k^2 \kappa \delta T$$

it follows from equations (10.10) that

$$\delta p = \frac{\omega^2}{k^2} \delta\rho \qquad -i\omega\rho T \frac{R}{\mu(\gamma - 1)} \left(\frac{\delta p}{p} - \gamma \frac{\delta\rho}{\rho} \right) = -k^2 \kappa \delta T.$$

† See, for instance, Landau L D and Lifshitz E M *Course of Theoretical Physics* vol 5 (Oxford: Pergamon) part 1, §43.

Together with the equation of state for an ideal gas: $p = R\rho T/\mu$, it gives the following dispersion relation $\omega(k)$ for the sound wave:

$$\frac{\omega^2}{k^2 c_0^2} = 1 - \frac{k^2\kappa(\gamma - 1)/\gamma}{k^2\kappa - \mathrm{i}\omega\rho R/\mu(\gamma - 1)} \qquad c_0^2 = \gamma\frac{p}{\rho}. \tag{10.11}$$

It is worth noting that in fact it is not necessary to solve the third-order equation (10.11) to obtain the relation $\omega(k)$. The reason is that within the field of application of the macroscopic gas dynamical equations the damping of the sound wave in a gas is usually weak (see below). Therefore ω may be represented in the form: $\omega \approx kc_0 - \mathrm{i}\Gamma$, with the damping rate $\Gamma \ll kc_0$. Correspondingly, the heat conductivity κ may be considered as a small quantity in the second term of the right-hand side in (10.11), so it may be omitted in the denominator, and we can put $\omega \approx kc_0$ there. It then follows at once that $\Gamma \approx (\gamma - 1)^2\mu k^2\kappa/2\gamma\rho R$. Let us adduce now some estimates showing that such a damping is really weak. In a gas the heat conductivity coefficient is usually of the order of magnitude of $\kappa \sim R\rho v_{\mathrm{T}} l/\mu$, where l is the mean free path of the molecules, and v_{T} is their mean thermal velocity. So the ratio $\Gamma/kc_0 \sim v_{\mathrm{T}} kl/c_0 \sim v_{\mathrm{T}} l/c_0\lambda$ (λ is the wavelength of the sound wave). Since the sound velocity in the gas is of the same order of magnitude as the thermal velocity of the molecules ($c_0 \sim v_{\mathrm{T}}$), the ratio $\Gamma/kc_0 \sim l/\lambda$, and is supposed to be small from the very meaning of the macroscopic description.

Problem 85

Determine the damping rate for a sound wave in a fluid arising from the presence of a small fraction of gas bubbles.

Let us consider a fluid with gas bubbles distributed in space chaotically but on average uniformly. They may be described by the distribution function over their size (radius) $f(r)$, so that $f(r)\,\mathrm{d}r$ is the number of bubbles per unit volume with the radius inside the interval $(r, r + \mathrm{d}r)$. We shall assume that their typical radius R,

$$R = \int_0^\infty fr\,\mathrm{d}r \Big/ \int_0^\infty f\,\mathrm{d}r$$

is small compared with the mean distance l between the bubbles ($l = (\int_0^\infty f\,\mathrm{d}r)^{-1/3}$). When the sound wave with the wavelength $\lambda \gg l$ propagates in this substance, the latter may be considered as a continuous medium, so the macroscopic equations for the averaged (over the distances much less than λ but large compared with l) quantities will be derived. Under this approach the linearized equations of the continuity and of the motion for the plane sound wave of small amplitude take the form

$$\frac{\partial}{\partial t}\langle \delta\rho\rangle + \langle \rho\rangle \frac{\partial v}{\partial x} = 0 \qquad \langle \rho\rangle \frac{\partial v}{\partial t} = -\frac{\partial}{\partial x}\langle \delta p\rangle$$

or

$$\frac{\partial^2}{\partial t^2}\langle \delta\rho\rangle = \frac{\partial^2}{\partial x^2}\langle \delta p\rangle$$

where $\langle \delta\rho\rangle$ and $\langle \delta p\rangle$ are the density and the pressure perturbations averaged in the above sense, while $\langle \rho\rangle$ is the mean density of this medium, which is equal to $\rho_0(1-\alpha)$. Here ρ_0 is the fluid density, and

$$\alpha = \frac{4\pi}{3}\int_0^\infty fr^3\,\mathrm{d}r \sim \left(\frac{R}{l}\right)^3 \ll 1$$

is the volume fraction occupied by the bubbles (the mass of the gas is negligible). To obtain the dispersion relation $\omega(k)$ (and so, the damping rate) we need to know the relation between $\langle \delta\rho\rangle$ and $\langle \delta p\rangle$. When the bubbles are absent we have simply $\delta\rho = \delta p/c_0^2$, where c_0 is the sound velocity in a pure fluid. However, now the pressure perturbations result in the forced oscillations of the bubbles, so a variation of the bubble radius ξ_r occurs (since for a given δp this variation depends on the bubble radius r, we have added to ξ a subscript r). Therefore, for a small value of α we have

$$\langle \delta\rho\rangle = \frac{\langle \delta p\rangle}{c_0^2} - 4\pi\rho_0\int_0^\infty f(r)r^2\xi_r\,\mathrm{d}r$$

where the value of ξ_r is described by the equation that follows from problem 70:

$$\frac{\mathrm{d}^2\xi_r}{\mathrm{d}t^2} + \omega_0^2(r)\xi_r = -\frac{\langle \delta p\rangle}{r\rho_0}$$

($\omega_0^2(r) = 3\gamma p/\rho r^2$ is the eigenfrequency of the radial oscillation of the bubble with radius r). For the perturbation of the form proportional to $\exp(ikx - i\omega t)$,

$$\xi_r = -\frac{\langle \delta p\rangle}{r\rho_0[\omega_0^2(r)-\omega^2]} \qquad \langle \delta\rho\rangle = \langle \delta p\rangle\left(c_0^{-2} + 4\pi\int_0^\infty \frac{f(r)r\,\mathrm{d}r}{\omega_0^2(r)-\omega^2}\right)$$

and the dispersion relation takes the form:

$$\omega^2 = k^2c_0^2\left(1 + 4\pi c_0^2\int_0^\infty \frac{f(r)r\,\mathrm{d}r}{\omega_0^2(r)-\omega^2}\right)^{-1}.$$

Without bubbles $\omega = kc_0$, so for a small α we may use a perturbation approach and put $\omega \approx kc_0$ in the denominator of the expression given above as the first approximation, then

$$\omega \approx kc_0\left(1 - 2\pi c_0^2 \int_0^\infty \frac{f(r) r\, \mathrm{d}r}{\omega_0^2(r) - k^2 c_0^2}\right) \tag{10.12}$$

Let us estimate the resulting correction to the wave frequency, $\Delta\omega$, assuming at first that there is no singularity resulting from the denominator in (10.12). In this case

$$\int_0^\infty \frac{f(r) r\, \mathrm{d}r}{\omega_0^2(r) - k^2 c_0^2} \sim \frac{\alpha}{R^2 \omega_0^2(R)} \sim \frac{\alpha \rho_0}{p}$$

so

$$\frac{\Delta\omega}{\omega} \sim \alpha \frac{c_0^2 \rho_0}{p}.$$

Because of the weak compressibility of a fluid the sound velocity $c_0 \gg (p/\rho_0)^{1/2}$ (this inequality becomes evident after taking into account the fact that $(p/\rho_0)^{1/2}$ is of the same order of magnitude as the sound velocity in a gas with the same pressure and density). Thus the correction is small if $\alpha \ll \varepsilon$, where $\varepsilon \equiv p/\rho_0 c_0^2 \ll 1$. However, even under this condition the presence of gas bubbles in the fluid may lead to a qualitatively new effect—the damping of the sound wave, if the denominator in the integral in (10.12) becomes equal to zero, that is bubbles with the eigenfrequency $\omega_0(r)$ equal to the frequency of the sound wave, are present in a fluid.

Such a resonant damping, which occurs in the absence of any dissipation processes is called *Landau damping* (it was first discovered by L D Landau for the Langmuir oscillations in a collisionless plasma). The nature of the singularity in the integral (10.12) is that ξ_r formally tends to infinity for the resonant bubbles. Therefore, we may avoid the singularity by taking into account an infinitesimal damping in the bubble eigenmode (for instance, because of the dissipation in a gas or in a fluid viscosity (see problem 70)). Then the eigenfrequency $\omega_0(r)$ acquires a small imaginary correction: $\omega_0(r) = \omega_0 - \mathrm{i}\nu$, $(\nu > 0)$, and the integral in (10.12) becomes as follows:

$$\int_0^\infty \frac{f(r) r\, \mathrm{d}r}{(\omega_0 + kc_0)(\omega_0 - kc_0 - \mathrm{i}\nu)} = \int_0^\infty \frac{f(r) r\, \mathrm{d}r}{(\omega_0 + kc_0)} \frac{(\omega - kc_0) + \mathrm{i}\nu}{(\omega - kc_0)^2 + \nu^2}.$$

Since

$$\lim_{\nu \to 0} \frac{\nu}{(\omega_0 - kc_0)^2 + \nu^2} = \pi\delta(\omega_0 - kc_0)$$

$$\mathrm{Im}\int_0^\infty \frac{f(r)r\,\mathrm{d}r}{(\omega + kc_0)(\omega_0 - kc_0 - \mathrm{i}\nu)} \approx \pi \int_0^\infty \frac{f(r)r\,\mathrm{d}r}{2\omega_0(r)}\,\delta(\omega_0(r) - kc_0\Big).$$

It is now seen that the introduction of the infinitesimal damping, which has no influence on the final result in the limit $\nu \to 0$, gives only the recipe of how to operate with the singular denominator in the integral. For $\nu > 0$ the pole is shifted down (since $\mathrm{d}\omega_0/\mathrm{d}r < 0$), so it needs to be passed above, and the result is reduced to the calculation of the half-residue at the point $\omega_0(r) = kc_0$. This is called the *Landau passing rule*. It follows now from (10.12) after the simple expansion that the damping decrement is equal to

$$\Gamma = \pi^2 c_0^2 \int_0^\infty f(r)r\,\mathrm{d}r\,\delta(\omega_0(r) - kc_0) = \frac{\pi^2 c_0^2 f(r_0) r_0}{|\mathrm{d}\omega_0/\mathrm{d}r_0|_{r_0}}$$

$$\omega_0(r_0) = kc_0.$$

If the distribution function of the bubbles, $f(r)$, is a smooth function with a typical scale of variation of the same order of magnitude as the radius of the resonant bubbles r_0, a simple estimate shows that $\Gamma \sim \alpha c_0^2/r_0^2\omega_0(r_0)$, and the ratio $\Gamma/\omega \sim \alpha c_0^2/r_0^2\omega_0^2(r_0) \sim \alpha/\varepsilon$. Therefore all the results obtained above by the perturbation method are valid only under a stronger restriction than $\alpha \ll 1$: it requires that $\alpha \ll \varepsilon \ll 1$. Otherwise, the presence of gas bubbles in a fluid leads to an appreciable change of the dispersion of sound waves in such a medium.

Chapter 11

Shock Waves

Recommended textbooks:
Landau L D and Lifshitz E M *Fluid Mechanics* (Oxford: Pergamon)
Whitham G B *Linear and Nonlinear Waves* (New York: Wiley)

Problem 86

In an ideal gas with pressure p_0, density ρ_0 and adiabatic specific ratio γ, an initial perturbation is created as the density variation $\delta\rho(x_0)$, which is equal to

$$\delta\rho = \begin{cases} \rho_1 \sin \pi x_0/L & |x_0| \leqslant L \\ 0 & |x_0| \geqslant L. \end{cases}$$

with no initial motion.

Determine the moment when a discontinuity (a shock wave) is formed in such a gas, considering the initial perturbation as weak ($\rho_1 \ll \rho_0$).

In the linear (acoustic) approximation such a perturbation after the time interval $\tau_0 \sim L/c_0$ will decay into two sound pulses, separated in space, and propagating in the right-hand and left-hand directions with constant profiles of the velocity, density and pressure (see problem 79). The steepening of the profiles of the pulses and the discontinuity formation arising from the non-linearity will occur during the time interval $\tau_1 \gg \tau_0$ (since for $\rho_1 \ll \rho_0$ the non-linearity is weak). Therefore, the non-linear evolution of the pulses may be considered independently of each other. According to the solution obtained in problem 79, in a linear approximation

$$\begin{aligned} v_+(x,t) &= \frac{c_0}{2}\frac{\rho_1}{\rho_0}\sin\frac{\pi}{L}(x - c_0 t) \\ v_-(x,t) &= -\frac{c_0}{2}\frac{\rho_1}{\rho_0}\sin\frac{\pi}{L}(x + c_0 t) \end{aligned} \qquad |x \pm c_0 t| \leqslant L. \tag{11.1}$$

Here v_+ and v_- correspond to the pulses propagating to the right and to the left, respectively, and expression (10.6) is used with $u(x_0) = 0$ and $a(x_0) = c_0^2\delta\rho = c_0^2\rho_1 \sin \pi x_0/L$. The distortion of the profile occurs because of the fact that the portions of the wave with different amplitudes propagate with velocities that differ from the sound velocity c_0 in an unperturbed gas. The exact non-linear solution for a *simple wave* (a Riemann wave) shows that points where the gas velocity has a given value move with constant speed

$$u(v) = v \pm c(v) \tag{11.2}$$

where $c(v)$ is the local value of the sound velocity (the two signs in (11.2) correspond to waves propagating to the right and to the left, respectively)†. Since in our case the non-linearity is weak ($\rho_1 \ll \rho_0$), the perturbation method may be used to obtain the value of $c(v)$. In a linear approximation it follows from the continuity equation that the pulses (11.1) are accompanied by the following density and pressure perturbations:

$$\delta\rho_\pm = \pm\frac{\rho_0}{c_0}v_\pm = \frac{\rho_1}{2}\sin\frac{\pi}{L}(x \mp c_0 t)$$

$$\delta p_\pm \approx c_0^2\delta\rho_\pm .$$

Therefore, we have for the local sound speed $c(v)$ an expansion in the form

$$c = \left(\frac{\gamma p}{\rho}\right)^{1/2} \approx \left(\gamma\frac{\rho_0 + \delta p}{\rho_0 + \delta\rho}\right)^{1/2} \approx c_0\left(1 + \frac{1}{2}\frac{\delta p}{p_0} - \frac{1}{2}\frac{\delta\rho}{\rho_0}\right)$$
$$\approx c_0 \pm \tfrac{1}{2}(\gamma - 1)v$$

so the propagation velocity (11.2) is equal to

$$u(v) = \pm c_0 + \tfrac{1}{2}(\gamma + 1)v \tag{11.3}$$

(it is worth noting that for the particular case of an ideal gas with constant γ this expression, obtained here for a weak non-linearity, coincides with the exact formula). Knowing $u(v)$ we can easily find the instant of the formation of the discontinuity, since this problem becomes analogous to that considered in problem 50, where the discontinuity was formed in a dust cloud. The only difference is that unlike the dust, where each portion of a medium continues to move with its initial velocity, in a compressible gas the points with a given velocity move with the speed (11.3). Therefore

$$x(x_0, t) = x_0 + u(v)t = x_0 \pm c_0 t + \tfrac{1}{2}(\gamma + 1)v_\pm t$$
$$= x_0 \pm c_0 t \pm \left(\frac{\gamma + 1}{2}\right)\frac{c_0 t}{2}\frac{\rho_1}{\rho_0}\sin\frac{\pi x_0}{L}. \tag{11.4}$$

† See, for instance, Landau LD and Lifshitz E M *Course of Theoretical Physics* vol 6 (Oxford: Pergamon) ch 10.

The discontinuity is formed when the derivative $(\partial x/\partial x_0)_t$ tends to zero. As follows from (11.4)

$$\left(\frac{\partial x}{\partial x_0}\right)_t = 1 \pm \frac{(\gamma + 1)c_0 t}{2} \frac{\rho_1}{2} \frac{\pi}{\rho_0 L} \cos \frac{\pi x_0}{L}. \qquad (11.5)$$

It is now seen that for a pulse moving to the right (the plus sign in (11.5)) the discontinuities form, first, at the points $x_0 = \pm L$, and occurs at the instant

$$\tau_1 = \frac{L}{c_0} \frac{4}{\pi(\gamma + 1)} \frac{\rho_0}{\rho_1} \gg \tau_0.$$

For a pulse propagating to the left, the discontinuity occurs at the same instant, but at the middle point of the pulse ($x_0 = 0$).

Problem 87

The shock wave with the Mach number $\mathcal{M}$ propagates in a gas with pressure p_1, density ρ_1 and adiabatic specific ratio γ. Determine the density and the pressure of this gas behind the shock wave.

The states of the gas in front of the shock wave and behind it (we shall denote the corresponding quantities by the subscripts 1 and 2) are related to each other by the following conservation laws: the continuity of the fluxes of the substance, of the momentum fluxes, and of the energy fluxes at the boundary of the discontinuity (see formulae (7.6) and (7.7)). So we obtain the following relations in the frame of reference in which the shock wave is at rest:

$$\begin{aligned} \rho_1 v_1 &= \rho_2 v_2 \\ p_1 + \rho_1 v_1^2 &= p_2 + \rho_2 v_2^2 \qquad (11.6) \\ w_1 + v_1^2/2 &= w_2 + v_2^2/2. \end{aligned}$$

For an ideal gas with constant γ the heat function is $w = \gamma p/\rho(\gamma - 1)$. Taking into account the fact that the Mach number of a shock wave is designated as the ratio of the shock wave speed and the sound speed in an unperturbed gas, so $v_1 = \mathcal{M} c_1$ (where $c_1^2 = \gamma p_1/\rho_1$), and introducing the dimensionless quantities $x = \rho_2/\rho_1$ and $y = p_2/p_1$, we obtain from (11.6):

$$1 + \gamma \mathcal{M}^2 = y + \gamma \mathcal{M}^2/x$$

$$(\gamma - 1)^{-1} + \mathcal{M}^2/2 = y/x(\gamma - 1) + \mathcal{M}^2/2x^2.$$

Eliminating the trivial solution ($x = y = 1$) we obtain that

$$\rho_2/\rho_1 = \frac{[(\gamma+1)\mathcal{M}^2]}{[2+(\gamma-1)\mathcal{M}^2]}$$
$$p_2/p_1 = (2\gamma\mathcal{M}^2 - \gamma + 1)/(\gamma+1). \tag{11.7}$$

As is seen from (11.7), for weak shock waves, when $\mathcal{M} - 1 \ll 1$,

$$\rho_2/\rho_1 \approx 1 + 4(\mathcal{M} - 1)/(\gamma + 1)$$
$$p_2/p_1 \approx 1 + 4\gamma(\mathcal{M} - 1)/(\gamma + 1).$$

This means that in this case $\delta p/p \approx \gamma\delta\rho/\rho$, so these quantities vary approximately by the adiabatic law. This shows that for a weak shock wave the change in the gas entropy is of a higher order in the small parameter ($\mathcal{M} - 1$).

In the opposite limit of strong shock waves, when the Mach number $\mathcal{M} \gg 1$, it follows from (11.7) that

$$\rho_2/\rho_1 \approx (\gamma + 1)/(\gamma - 1)$$
$$p_2/p_1 \approx 2\gamma\mathcal{M}^2/(\gamma + 1). \tag{11.8}$$

Therefore, behind the strong shock wave the degree of compression of the gas remains finite, so the gas pressure grows mainly because of the strong heating of the gas.

Problem 88

A strong shock wave propagating in a polytropic gas ($\gamma = \text{const}$) with the Mach number $\mathcal{M}_0 \gg 1$, is reflected from a plane rigid wall (figure 11.1). Derive the Mach number $\mathcal{M}_1$ for the reflected shock wave.

In the reference frame with a gas at rest near the wall, the incident shock

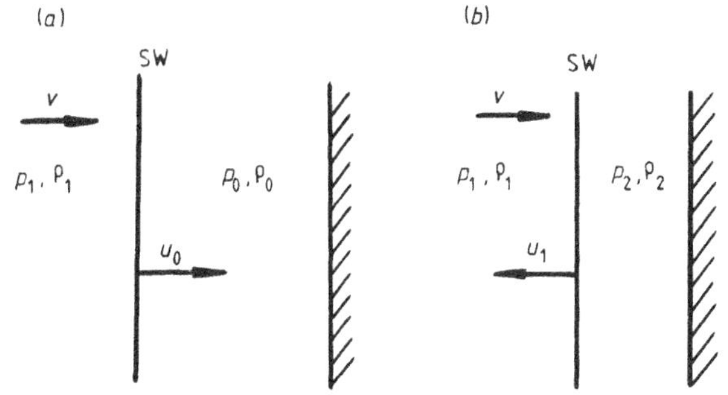

Figure 11.1. Reflection of a shock wave (sw).

wave moves with the velocity $u_0 = \mathcal{M}_0 c_0$, while a gas behind it has the velocity v (figure 11.1(a)). For this strong shock wave the relations (11.8) may be applied, so the first of them gives:

$$\rho_1/\rho_0 = u_0/(u_0 - v) \approx (\gamma + 1)/(\gamma - 1) \qquad (11.9a)$$

that is

$$v \approx 2\mathcal{M}_0 c_0/(\gamma + 1) \qquad (11.9b)$$

The reflected shock wave (figure 11.1(b)) moves relative to the gas in front of it with the velocity $u_1 + v$, so that $u_1 + v = \mathcal{M}_1 c_1$, where c_1 is the speed of sound in this gas. It follows from (11.8) that

$$c_1 = (\gamma p_1/\rho_1)^{1/2} \approx c_0 \frac{\mathcal{M}_0}{\gamma + 1}[2\gamma(\gamma - 1)]^{1/2}. \qquad (11.10)$$

After this we may write for the degree of compression in the reflected shock wave, according to (11.7):

$$\frac{\rho_2}{\rho_1} = \frac{(u_1 + v)}{u_1} = \frac{\mathcal{M}_1 c_1}{(\mathcal{M}_1 c_1 - v)} = \frac{(\gamma + 1)\mathcal{M}_1^2}{2 + (\gamma - 1)\mathcal{M}_1^2}.$$

By substitution of the expressions (11.9) and (11.10) for v and c_1 we obtain the following equation for $\mathcal{M}_1$:

$$\mathcal{M}_1^2 - \frac{(\gamma + 1)}{[2\gamma(\gamma - 1)]^{1/2}}\mathcal{M}_1 - 1 = 0$$

with the appropriate solution (the second root is negative):

$$\mathcal{M}_1 \approx 2\gamma/(\gamma - 1)^{1/2}.$$

It is seen that the reflected shock wave has a Mach number of the order of unity, so the intensity of the shock wave is appreciably reduced under the reflection.

Problem 89

Determine the internal structure of a weak shock wave in a viscous gas.

As follows from the results obtained in problem 86, a simple wave in an ideal gas may be described by the following equation:

$$\frac{\partial v}{\partial t} + \left(c_0 + \frac{\gamma + 1}{2}v\right)\frac{\partial v}{\partial x} = 0. \qquad (11.11)$$

During its evolution a steepening of the wave profile takes place, so the wave finally breaks up. However, in a real gas even small dissipative processes (for instance, viscosity or heat conductivity) come into play while the wave profile becomes sufficiently steep. Dissipation prevents the breaking up of the wave and results in the formation of a shock wave of finite width. To describe this structure let us add the viscous term to the equation of motion (11.11)

$$\frac{\partial v}{\partial t} + \left(c_0 + \frac{\gamma + 1}{2} v\right)\frac{\partial v}{\partial x} = \nu \frac{\partial^2 v}{\partial x^2} \tag{11.12}$$

where ν is the kinematic viscosity of the gas. It must be noted that such a procedure is correct only for the description of a weak shock wave. The reason is that equation (11.11) has been derived for an ideal gas under the condition of adiabaticity. However, as was mentioned in problem 87, for a weak shock wave the entropy production is of higher order than the change in density or pressure. Therefore, the equation (11.12) is still valid for this case.

By the passage to the reference frame moving with the sound speed c_0, and the introduction of a new function $u = \frac{1}{2}(\gamma + 1)v$, the equation (11.12) can be brought to the form

$$\frac{\partial u}{\partial t} + u\frac{\partial u}{\partial x} = \nu \frac{\partial^2 u}{\partial x^2} \tag{11.13}$$

known as the *Burgers equation*. Here we are interested in a particular solution of this equation, describing the structure of a steady shock wave, that propagates in a gas with a constant velocity. Therefore we search for the solution of (11.13) in the form

$$u = f(x - v_0 t) \tag{11.14}$$

so v_0 is the velocity of a shock wave in the chosen reference frame. By substitution of (11.14) into (11.13) we obtain the following equation for $f(z)$:

$$-v_0 \frac{df}{dz} + f\frac{df}{dz} = \nu \frac{d^2 f}{dz^2}.$$

After a simple integration over z this gives

$$-v_0 f + \tfrac{1}{2} f^2 = \nu \frac{df}{dz} + \text{const}$$

or

$$\nu \frac{df}{dz} = \tfrac{1}{2}(f - v_0)^2 + a$$

where a is a new constant of integration. For the solution of interest to us df/dz tends asymptotically to zero at $z \to \pm\infty$. Therefore, the constant a

must be negative: $a = -w_0^2/2$. Then

$$\nu \frac{df}{dz} = \frac{(f - v_0)^2}{2} - \frac{w_0^2}{2}$$

so the asymptotic values of f as $z \to \pm\infty$ are, respectively, $f_+ = v_0 + w_0$, and $f_- = v_0 - w_0$. For a new function $\varphi(z)$, defined as $\varphi(z) = (f - v_0)/w_0$, we now obtain the following equation:

$$\varphi^2 - 1 = \frac{2\nu}{w_0} \frac{d\varphi}{dz}$$

and thus

$$\varphi(z) = \tanh(w_0 z/2\nu).$$

In a weak shock wave the discontinuity in the gas velocity, $v_1 - v_2$, is equal, according to (11.7), to

$$v_1 - v_2 \approx 4c_0(\mathcal{M} - 1)/(\gamma + 1).$$

On the other hand, in the notation used above, $v_1 - v_2 = 4w_0/(\gamma + 1)$. Therefore, it follows from (11.15) that the thickness of the weak shock wave is

$$\delta = 2\nu/w_0 = 2\nu/c_0(\mathcal{M} - 1).$$

Since the kinematic viscosity of a gas is usually of the order of $\nu \sim v_T l$, where l is the mean free path of the gas molecules, the weak shock wave thickness is $\delta \sim l/(\mathcal{M} - 1)$. Thus the thickness of a sufficiently strong shock wave (when $(\mathcal{M} - 1) \gtrsim 1$) becomes of the same order of magnitude as l, so its structure cannot be described by the macroscopic equations.

Chapter 12

Non-linear Waves in Dispersive Media. Solitons

Recommended textbook:
Whitham G B *Linear and Nonlinear Waves* (New York: Wiley)

Problem 90

Derive the dispersion law for a gravitational wave propagating on the free surface of an incompressible fluid with the finite depth h_0.

We shall consider (as in problem 55) that in the unperturbed state the fluid surface is the horizontal plane $z = 0$, and that a wave propagates along the x-axis. Unlike the case of an infinitely deep fluid, in the general solution of the Laplace equation for the velocity potential

$$\varphi(x, z, t) = (B_1\, e^{kz} + B_2\, e^{-kz})\, e^{i(kx-\omega t)}$$

the constant B_2 now differs from zero and has to be obtained from the condition that the normal component of the fluid velocity vanishes at the bottom:

$$v_z|_{z=-h_0} = -\left.\frac{\partial\varphi}{\partial z}\right|_{z=-h_0} = 0$$

so that

$$\varphi = B\cosh[k(z+h_0)]\, e^{i(kx-\omega t)}. \qquad (12.1)$$

It then follows from the linearized equation of motion (7.17) that a pressure distribution in a fluid becomes equal to

$$p = p_0 - \rho g z + \delta p = p_0 - \rho g z - i\omega\rho\varphi$$
$$= p_0 - \rho g z - i\omega B\rho\cosh[k(z+h_0)]\, e^{i(kx-\omega t)}. \qquad (12.2)$$

Denoting the vertical displacement of the boundary surface by

$\xi(x, t) = \xi_0 \exp[i(kx - \omega t)]$, we may relate the constants B and ξ_0 by the condition

$$\partial\xi/\partial t = -\partial\varphi/\partial z|_{z=0}$$

so that $B = i\omega\xi_0/k \sinh kh_0$. The dispersion relation for such a gravitational wave now follows from the boundary condition on the fluid surface:

$$p|_{z=\xi} = p_0. \tag{12.3}$$

With the accuracy needed for a linear wave we obtain from (12.2) and (12.3) that

$$\omega(k) = (kg \tanh kh_0)^{1/2}. \tag{12.4}$$

Thus it is seen that a finite value for the fluid depth becomes essential for motions of sufficiently long wavelength, when $\lambda = 2\pi/k \gtrsim h_0$. For short-wavelength perturbations ($\lambda \ll h_0$) we come back to the previous result found in problem 55 for an infinitely deep fluid: $\omega = (gk)^{1/2}$. In the opposite limit, when $\lambda \gg h_0$ (shallow water) it follows from (12.4) that

$$\omega \approx \pm u_0 k \qquad u_0 = (gh_0)^{1/2} \tag{12.5}$$

so such a wave propagates along the fluid surface with velocity equal to u_0.

Problem 91

Derive the equation for a shallow-water gravitational wave ($\lambda \gg h_0$), taking into account the weak non-linearity and dispersion.

In considering equations for the shallow-water waves it is convenient to take into account at once the simplifications arising in this limit. Thus, it follows from expressions (12.1) and (12.2) that the inequality $v_x \gg v_z$ is now valid, and v_x, as well as the pressure perturbation δp, has a weak dependence on z. Therefore it becomes possible to consider the problem as one dimensional in the first approximation, so that $v_x \equiv v$ and δp are functions only of the x-coordinate. Since a vertical displacement of the fluid surface $\xi(x, t)$ arises while the wave propagates, the net depth of the fluid at a given point becomes equal to $h = h_0 + \xi(x, t)$. Then the mass conservation law for an incompressible fluid (in analogy with the continuity equation) takes the following form:

$$\frac{\partial h}{\partial t} + \frac{\partial}{\partial x}(hv) = 0. \tag{12.6}$$

In the equation of motion

$$\frac{\partial v}{\partial t} + v\frac{\partial v}{\partial x} = -\frac{1}{\rho}\frac{\partial \delta p}{\partial x}$$

the pressure perturbation δp may be expressed over the surface displacement ξ by taking into account equation (12.2) for the net pressure and the boundary condition (12.3), which gives $\delta p = -\rho g \xi$, so that

$$\frac{\partial v}{\partial t} + v\frac{\partial v}{\partial x} = -g\frac{\partial \xi}{\partial x} = -g\frac{\partial h}{\partial x}. \tag{12.7}$$

In a linear approximation equations (12.6) and (12.7) result in

$$\frac{\partial h}{\partial t} + h_0\frac{\partial v}{\partial x} = 0 \qquad \frac{\partial v}{\partial t} + g\frac{\partial h}{\partial x} = 0$$

which leads to the wave equation

$$\frac{\partial^2 h}{\partial t^2} - gh_0\frac{\partial^2 h}{\partial x^2} = 0$$

with the dispersion law (12.5) obtained above. Now we shall derive the exact solution of the non-linear equations (12.6) and (12.7) in the form of a simple wave (see problem 86). In this case a one-to-one relation between $h(x, t)$ and $v(x, t)$ is assumed, so that $h(x, t) \equiv h(v)$, and then it follows from (12.6) and (12.7) that

$$\frac{\mathrm{d}h}{\mathrm{d}v}\left(\frac{\partial v}{\partial t} + v\frac{\partial v}{\partial x}\right) = -h\frac{\partial v}{\partial x} \qquad \frac{\partial v}{\partial t} + v\frac{\partial v}{\partial x} = -g\frac{\partial v}{\partial x}\frac{\mathrm{d}h}{\mathrm{d}v}. \tag{12.8}$$

By division of the right- and left-hand sides of these relations by each other we obtain that

$$(\mathrm{d}h/\mathrm{d}v)^2 = h/g \tag{12.9a}$$

that is

$$\mathrm{d}h/\mathrm{d}v = \pm(h/g)^{1/2} \tag{12.9b}$$

so the second of equations (12.8) takes the form

$$\frac{\partial v}{\partial t} + [v \pm (gh)^{1/2}]\frac{\partial v}{\partial x} = 0. \tag{12.10}$$

This means that a point with a given value of the fluid velocity v (and, hence, h) moves with the velocity

$$u = v \pm (gh)^{1/2}. \tag{12.11}$$

Thus this solution is analogous to the Riemann wave in gas dynamics (see formula (11.2)) with the quantity $(gh)^{1/2}$ playing the role of the sound velocity (the two signs correspond to waves travelling to the right and left along the x-axis). In a linear approximation, $u = \pm(gh_0)^{1/2}$. For the case of weak non-linearity we may obtain from (12.11) that

$$u \approx v \pm (gh_0)^{1/2} \pm \tfrac{1}{2}(g/h_0)^{1/2}(h - h_0)$$

and since, according to (12.9),

$$h = h_0 + \frac{dh}{dv}\bigg|_{h=h_0} v = h_0 \pm \left(\frac{h_0}{g}\right)^{1/2} v$$

we have

$$u \approx \tfrac{3}{2}v \pm (gh_0)^{1/2}. \tag{12.12}$$

It is now seen by comparing expressions (12.12) and (11.3) that in a one-dimensional approximation the shallow-water waves become completely analogous to the perturbations in an ideal gas with the adiabatic specific ratio $\gamma = 2$. Therefore the steepening of the wave profile mentioned above (see problem 86) will occur while a wave propagates. In a gas this process continues until the dissipation (viscosity and thermal conductivity) comes into play (see problem 89) and a shock wave is formed. However, for the shallow water the non-linear steepening of the gravitational wave is stopped much earlier owing to the *dispersion effects*. As has been mentioned above, equations (12.6) and (12.7) represent the first-order approximation in the small parameter h_0/λ. Since the steepening of the wave results in the effective reduction of λ (short-wavelength harmonics are formed), the effects of the higher order in h_0/λ come into play. As will be shown below, they lead to the fact that the propagation velocity of a wave differs slightly from $u_0 = (gh_0)^{1/2}$ and, what is the important point, the propagation velocity becomes dependent on the harmonic wavelength (dispersion arises). The latter results in the spread of the wave packet (see problem 30) and therefore impedes the non-linear steepening. Thus, a competition between these processes leads to the formation of non-linear structures with a steady profile propagating on the shallow-water surface (see the next problem).

Since the weak non-linear and dispersive effects may be analysed separately from each other, we shall derive the latter from the linear dispersion equation (12.4) by keeping the term of the next order in a small parameter kh_0. Then we obtain, instead of (12.5), the following result:

$$\omega \approx \pm k u_0 (1 - k^2 h_0^2/6). \tag{12.13}$$

Now the problem is to take into account this dispersion effect in the equation (12.10) for a simple wave, written in the (x, t) representation. This may be done by adding to equation (12.10) such a term that gives the required correction (12.13) to the dispersion law. It is easy to see that for a wave travelling to the right, this equation must become as follows:

$$\frac{\partial v}{\partial t} + u_0 \frac{\partial v}{\partial x} + \frac{u_0 h_0^2}{6} \frac{\partial^3 v}{\partial x^3} = 0.$$

Contributing also a weak non-linearity with the help of expression (12.12), we obtain

$$\frac{\partial v}{\partial t} + u_0 \frac{\partial v}{\partial x} + \frac{3}{2} v \frac{\partial v}{\partial x} + \frac{u_0 h_0^2}{6} \frac{\partial^3 v}{\partial x^3} = 0. \tag{12.14}$$

The second term of this equation has no effect on the wave profile evolution and describes the motion of a perturbation as a whole with a velocity u_0. So after the transition to the accompanying frame of reference and the introduction of a new function $V = \frac{3}{2}v$, we finally obtain the following equation

$$\frac{\partial V}{\partial t} + V \frac{\partial V}{\partial x} + \beta \frac{\partial^3 V}{\partial x^3} = 0 \qquad \beta = \tfrac{1}{6} u_0 h_0^2 > 0 \tag{12.15}$$

known as the *Korteweg–de Vries equation*. It has been intensively studied during the last two decades, since many problems in the physics of non-linear waves in dispersive media may be described by this universal equation (such a reduction may be achieved by the appropriate transformation of the variables in the particular physical problem). The remarkable new method of solving the non-linear evolution equations—the inverse scattering transform, was first discovered while considering the Korteweg–de Vries equation†.

Problem 92

Derive a shape of the steady non-linear waves described by the Korteweg–de Vries equation.

Let us consider the solution of equation (12.15) in the form of a travelling wave with the steady shape:

$$V = f(x - ut) \tag{12.16}$$

(remembering that equation (12.15) has been derived in a reference frame moving with a velocity u_0, so the net wave speed is $u_0 + u$). By the straightforward substitution of (12.16) into (12.15) we get

$$-uf' + ff' + \beta f''' = 0$$

and after one integration,

$$f'' = \frac{u}{\beta} f - \frac{f^2}{2\beta} + C \tag{12.17}$$

where C is a constant of integration. It is easy to see that its value specifies

† See, for instance, Newell A C The inverse scattering transform *Topics in Current Physics* vol 17, eds R Bullough and P Caudrey (Berlin: Springer).

the frame of reference, so it is always possible to put $C = 0$ by the transition to some moving reference frame. Indeed, by adding the same constant to the velocities u and f:

$$\tilde{u} = u + a \qquad \tilde{f} = f + a \tag{12.18}$$

we obtain from (12.17) the following equation for $\tilde{f}$:

$$\tilde{f}'' = \frac{\tilde{u}}{\beta}\tilde{f} - \frac{\tilde{f}^2}{2\beta} + \left(\frac{a^2}{2\beta} - \frac{\tilde{u}a}{\beta} + C\right).$$

The expression in brackets may be turned to zero by the appropriate choice of a, so we return (in terms of $\tilde{f}$ and $\tilde{u}$) to the equation (12.17), but with $C = 0$. The latter may be analysed with the help of a mechanical analogy after writing (12.17) (with $C = 0$) in the form

$$\frac{\mathrm{d}^2 f}{\mathrm{d}x^2} = \frac{u}{\beta}f - \frac{f^2}{2\beta} \equiv \mathscr{F}(f). \tag{12.19}$$

By the interpretation of f as a coordinate, and of x as a time, this becomes the equation of motion for a point of unit mass under the force $\mathscr{F}(f)$, so the corresponding potential $W(f)$ is as follows:

$$\mathscr{F} = -\frac{\mathrm{d}W}{\mathrm{d}f} \qquad W(f) = -\frac{uf^2}{2\beta} + \frac{f^3}{6\beta}. \tag{12.20}$$

A plot of $W(f)$ is shown in figure 12.1 where $u < 0$ is considered. The 'finite trajectories', which are of interest to us, correspond to the energy interval $0 < E < 2|u|^3/3\beta$. For low energies a particle oscillates near the bottom of the potential well (point $f = 0$), where the shape of $W(f)$ is approximately parabolic (the second term in (12.20) is negligible). Therefore, we have here harmonic oscillations, so, according to (12.16),

$$V(x,t) \approx V_0 \exp[\mathrm{i}(|u|/\beta)^{1/2}(x - ut)]. \tag{12.21}$$

This wave corresponds to a linear approximation. In this case the shift of the phase velocity is negative ($u < 0$) (as is prescribed by equation (12.13), and is equal to $u = -k^2\beta$ ($k = (|u|/\beta)^{1/2}$ in the solution (12.21)). The role of non-linearity increases with the increase of energy, and the non-symmetry of the shape of $W(f)$ (see figure 12.1) becomes important: the growth is flatter for $f < 0$ and steeper for $f > 0$. Therefore, in the non-linear periodic wave the maxima of V become relatively narrow (a 'particle' passes over this interval of f quickly, while the minima are flat (here a 'particle' moves slowly). In the limit when $E = 2|u|^3/3\beta$, the left-hand reflection point ($f = -2|u|$) is reached by a 'particle' asymptotically at the 'time' $t \to \infty$ (the period of oscillation tends to infinity). Therefore, while x varies from $-\infty$ to $+\infty$, f grows from the value $-2|u|$ up to the maximum, equal to $|u|$, and then reduces back to $-2|u|$. This particular solution is called a *soliton* (or a solitary wave). If now according to the rule (12.18), we add

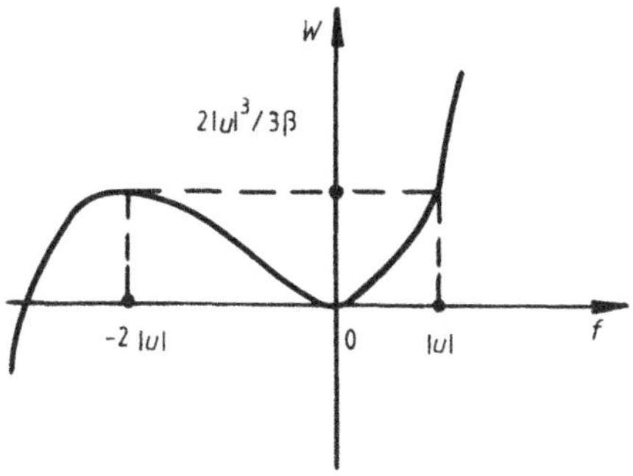

Figure 12.1. Effective potential energy for travelling non-linear waves.

the same velocity, $a = 2|u|$, to f and u, then in such a frame of reference the wave perturbation will be absent at infinity ($f|_{x \to \pm\infty} \to 0$), its maximum value becomes $f_{max} = 3|u|$, and the soliton will move to the right with the velocity $|u| = u$ (we now put $u > 0$). Thus, on the surface of a fluid, which is at rest at infinity, the soliton propagates with the 'hypersound' velocity, $u_0 + u$, with $u = \frac{1}{3} f_{max}$. Let us now determine the shape of a soliton by integration of equation (12.19):

$$\frac{1}{2}\left(\frac{df}{dx}\right)^2 + W(f) = E = \text{const.} \tag{12.22}$$

It is easy to see that in the reference frame used now, a 'particle' energy $E = 0$ corresponds to a soliton. So we may obtain from (12.22) and (12.20) that

$$df/dx = \pm(u/\beta)^{1/2} f(1 - f/3u)^{1/2}. \tag{12.23}$$

Let us position the peak of the soliton (the point, where $f = f_{max} = 3u$) at the coordinate point $x = 0$. Then in the region $x > 0$ the perturbation decreases (the negative sign in the expression (12.23)), so

$$\int_{f/f_{max}}^{1} \frac{dz}{z(1 - z)^{1/2}} = \left(\frac{u}{\beta}\right)^{1/2} x.$$

After a simple integration we have

$$\frac{1 + (1 - f/f_{max})^{1/2}}{(f/f_{max})^{1/2}} = \exp[(u/4\beta)^{1/2} x].$$

So, finally, the shape of the soliton is given by the expression

$$f(x) = 3u/\cosh^2(x/l) \qquad l = (4\beta/u)^{1/2}.$$

Thus, the characteristic width of the soliton, l, decreases with the growth of its amplitude, u, proportionally to $u^{-1/2}$. This result possesses a simple physical explanation. As has been mentioned above, a steady shape of the soliton is provided by the competition between the non-linearity (which leads to the steepening) and the dispersion (which results in the spreading). Comparing by the order of magnitude the non-linear and dispersive terms in the Korteweg–de Vries equation (12.15), we may now obtain the following:

$$V\frac{\partial V}{\partial x} \sim \frac{u^2}{l} \sim \beta\frac{\partial^3 V}{\partial x^3} \sim \beta\frac{u}{l^3}$$

that is

$$l \sim (\beta/u)^{1/2}.$$

Chapter 13

The Theory of Elasticity

Recommended textbooks:
Landau L D and Lifshitz E M *Course of Theoretical Physics, Theory of Elasticity* (Oxford: Pergamon)
Atkin R J and Fox N *An Introduction to the Theory of Elasticity* (London: Longman)

Small deformations of solid bodies are described by the *strain tensor*

$$u_{ik} = \frac{1}{2}\left(\frac{\partial u_i}{\partial x_k} + \frac{\partial u_k}{\partial x_i}\right) \tag{13.1}$$

where $\boldsymbol{u}$ is the *displacement vector*. Deformations are accompanied with internal forces characterized by the *stress tensor* σ_{ik}, so that the force acting on some portion of the body is

$$\mathscr{F}_i = \oint \sigma_{ik}\, \mathrm{d}S_k \tag{13.2}$$

where integration in (13.2) is carried out over the bounding surface, with the vector $\mathrm{d}\boldsymbol{S}$ directed along the external normal. Thus, the force per unit volume is equal to

$$f_i = \frac{\partial \sigma_{ik}}{\partial x_k} \qquad \text{so} \qquad \mathscr{F}_i = \int f_i\, \mathrm{d}V. \tag{13.3}$$

An isotropic elastic medium is usually characterized by either of two sets of elastic coefficients. This may be the *modulus of compression*, K, and the *modulus of rigidity*, μ, ($K \geqslant 0$, $\mu \geqslant 0$), or the *modulus of extension* (*Young's modulus*), E, and *Poisson's ratio*, σ. These two sets are related to each other as follows:

$$E = gK\mu/(3K+\mu) \qquad \sigma = \tfrac{1}{2}(3K-2\mu)/(3K+\mu). \tag{13.4}$$

It is seen from (13.4) that $E \geqslant 0$, and $-1 \leqslant \sigma \leqslant \frac{1}{2}$ (but there are no substances known for which σ is negative, so we shall consider in what follows that $0 \leqslant \sigma \leqslant \frac{1}{2}$).

For small deformations the stress tensor σ_{ik} is proportional to the strain tensor (*Hooke's law*). In an isotropic medium this relation takes the form:

$$\sigma_{ik} = Ku_{ll}\delta_{ik} + 2\mu(u_{ik} - \tfrac{1}{3}u_{ll}\delta_{ik})$$
$$u_{ik} = \frac{1}{gK}\sigma_{ll}\delta_{ik} + \frac{1}{2\mu}(\sigma_{ik} - \tfrac{1}{3}\sigma_{ll}\delta_{ik}) \qquad (13.5)$$

or, in terms of E and σ:

$$\sigma_{ik} = \frac{E}{1+\sigma}\left(u_{ik} + \frac{\sigma}{1-2\sigma}u_{ll}\delta_{ik}\right)$$
$$u_{ik} = \frac{1}{E}[(1+\sigma)\sigma_{ik} - \sigma\sigma_{ll}\delta_{ik}]. \qquad (13.6)$$

The free energy of a deformed elastic medium (per unit volume), $\mathcal{F}$, is equal to

$$\mathcal{F} = \sigma_{ik}u_{ik}/2 \qquad (13.7)$$

which, taking into account expressions (13.5) and (13.6), may be represented as

$$\mathcal{F} = \mu(u_{ik} - \tfrac{1}{3}u_{ll}\delta_{ik})^2 + \tfrac{1}{2}K(u_{ll})^2$$

or (13.8)

$$\mathcal{F} = \frac{E}{2(1+\sigma)}\left(u_{ik}^2 + \frac{\sigma}{1-2\sigma}u_{ll}^2\right).$$

The equation of equilibrium, according to (13.3), takes the form

$$\partial\sigma_{ik}/\partial x_k + \rho g_i = 0 \qquad (13.9)$$

where ρ is the density of the medium, and $\boldsymbol{g}$ is the external (non-elastic) force per unit mass. In terms of the displacement vector $\boldsymbol{u}$ the equilibrium equation (13.9) in an isotropic medium may be written as follows:

$$\nabla^2\boldsymbol{u} + \frac{1}{1-2\sigma}\nabla\,\mathrm{div}\,\boldsymbol{u} = -\rho\boldsymbol{g}\frac{2(1+\sigma)}{E}. \qquad (13.10)$$

Problem 93

A cube of edge l, made from an elastic material with given E and σ, is immersed in an absolutely rigid cavity of the same size (figure 13.1). Determine the deformation of the cube if a pressure p is applied on its free side.

Using the coordinate system indicated in the figure 13.1, we obtain at once that the displacement vector $\boldsymbol{u}$ has here only one non-zero component, u_z,

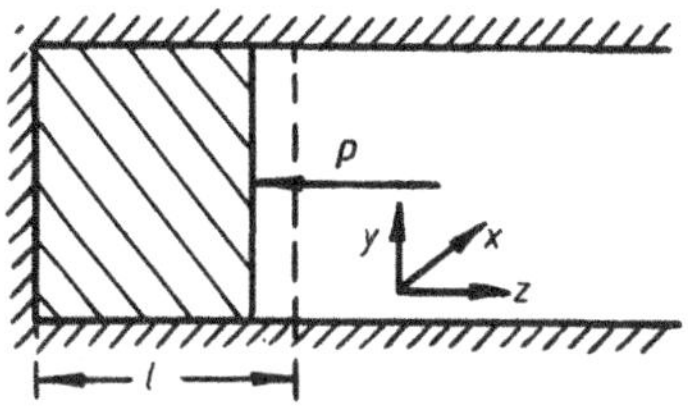

Figure 13.1 Compression of a cube in a rigid cavity.

which depends only on the z-coordinate: $u_z(z) \equiv u(z)$. Corresponding to this only the component $u_{zz} = du/dz$ is present in the strain tensor u_{ik} (see formula (13.1)). Then it follows from Hooke's law (13.6) for the stress tensor σ_{ik} that

$$\sigma_{xx} = \sigma_{yy} = \frac{E\sigma}{(1+\sigma)(1-2\sigma)} u_{zz} \qquad \sigma_{zz} = \frac{E(1-\sigma)}{(1+\sigma)(1-2\sigma)} u_{zz}. \tag{13.11}$$

The condition of equilibrium $\partial\sigma_{ik}/\partial x_k = 0$ results now in u_{zz} being a constant. The value of this uniform deformation has been determined from the boundary condition at the side $z = l$: $\sigma_{zz}(l) = -p$. Therefore, it follows from the second expression in (13.11) that

$$u_{zz} = -\frac{p(1+\sigma)(1-2\sigma)}{E(1-\sigma)}.$$

Problem 94

An isotropic elastic crystal occupies the half-space $z < 0$, and its boundary surface (the plane $z = 0$) is continuous with a perfectly rigid medium. Find the reflection coefficient for a longitudinal monochromatic sound wave incident at angle θ_l on the boundary surface (figure 13.2).

When the incident wave (wave 1) is reflected, there are in general both longitudinal (wave 2) and transverse (wave 3) reflected waves. Let us note their amplitudes of displacement as $\boldsymbol{u}_1$, $\boldsymbol{u}_2$ and $\boldsymbol{u}_3$ (the directions of these vectors is shown in the figure 13.2). For a given $\boldsymbol{u}_1$ the vectors $\boldsymbol{u}_2$ and $\boldsymbol{u}_3$ may be determined from a boundary condition. In this case the latter consists in the absence of the net displacement of a crystal at $z = 0$. Then we obtain the following equations:

$$u_z(0) = u_1 \cos\theta_l - u_2 \cos\theta_l + u_3 \sin\theta_t = 0$$

$$u_x(0) = u_1 \sin\theta_l + u_2 \sin\theta_l + u_3 \cos\theta_t = 0$$

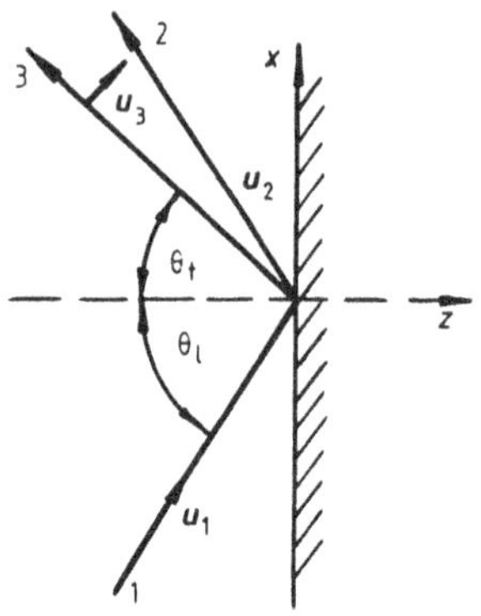

Figure 13.2 Reflection of an elastic wave from a rigid boundary.

so that

$$u_2 = u_1 \cos(\theta_l + \theta_t)/\cos(\theta_l - \theta_t)$$

and the reflection coefficient of interest is

$$R_l = \cos^2(\theta_l + \theta_t)/\cos^2(\theta_l - \theta_t). \qquad (13.12)$$

The angle θ_t of the reflected transverse wave can be expressed easily over the incident wave angle θ_l and the longitudinal and transverse velocities of sound, c_l and c_t. Since the frequencies of the longitudinal and transverse waves are the same, we have $c_l k_l = c_t k_t$, and both wave also have the same x-components of the wave vectors, so that

$$k_l \sin\theta_l = k_t \sin\theta_t.$$

Thus $\sin\theta_t = (c_t/c_l)\sin\theta_l$. It follows from the expression (13.12) that the reflection coefficient R_l tends to zero when $\theta_l + \theta_t = \frac{1}{2}\pi$. This result has an evident geometrical sense: in this case the vectors $\boldsymbol{u}_1$ and $\boldsymbol{u}_3$ become collinear, so the boundary condition $\boldsymbol{u}|_{z=0} = 0$ is satisfied at $\boldsymbol{u}_3 = -\boldsymbol{u}_1$ and $\boldsymbol{u}_2 = 0$.

Problem 95

A plug with length L and radius a, made from the elastic material with a given Young's modulus E and Poisson's ratio σ, is placed in a perfectly rigid cylindrical channel, so that the pressure p acts at the lateral surface of the plug. What is the minimum pushing force $\mathscr{F}_{\min}$ needed to move this plug (figure 13.3(a)), if the coefficient of friction between the plug and the surface of the channel is equal to K ($K \ll 1$)?

This problem becomes non-trivial because of the presence of Poisson's ratio

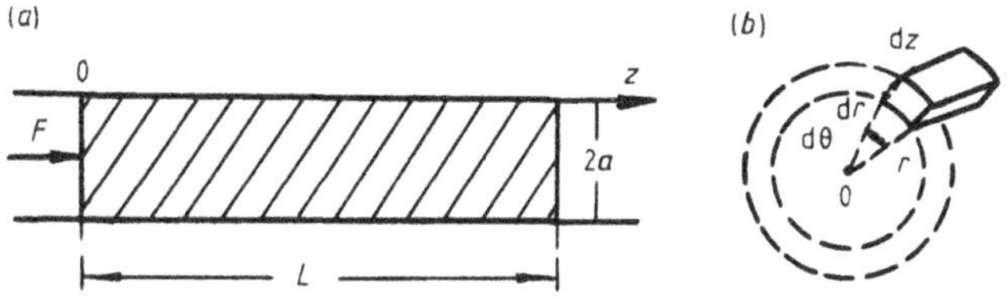

Figure 13.3 (*a*) Inserting a plug into a channel and (*b*) calculation of the equilibrium.

σ for the plug material. In the case when $\sigma = 0$ the pushing force has no influence on the cross stresses, so the pressure acting on the lateral surface of the plug remains equal to p. Therefore, the minimum pushing force in this case is $\mathscr{F}_{\min} = 2\pi a L K p$. However, for $\sigma > 0$ the pushing force results in an increase of the pressure, and it leads to a considerable growth of $\mathscr{F}_{\min}$ for a rather long plug ($L \gg a$).

Let us start with the equilibrium equations in cylindrical coordinates, which are natural for this problem. So we separate the elementary volume shown in figure 13.3(*b*), and sum all the forces acting on its surfaces. As a result we obtain that

$$\begin{gathered}\frac{1}{r}\frac{\partial}{\partial r}(r\sigma_{rr}) - \frac{\sigma_{\varphi\varphi}}{r} + \frac{\partial \sigma_{zr}}{\partial z} = 0 \\ \frac{1}{r}\frac{\partial}{\partial r}(r\sigma_{zr}) + \frac{\partial \sigma_{zz}}{\partial z} = 0\end{gathered} \tag{13.13}$$

(the azimuthal equilibrium is fulfilled automatically because of the azimuthal symmetry of our problem). In the absence of a pushing force $\mathscr{F}$ there are only two non-zero components of the stress tensor, σ_{rr} and $\sigma_{\varphi\varphi}$, so that

$$\sigma_{\varphi\varphi} = \sigma_{rr} = -p$$

(the first equality results from the radial equilibrium in (13.13), while the second one is the consequence of the boundary condition at the plug's lateral surface). In this case the plug is deformed uniformly, and as follows from Hooke's law (13.6)

$$u_{rr} = u_{\varphi\varphi} = -p(1-\sigma)/E \qquad u_{zz} = 2p\sigma/E. \tag{13.14}$$

Under the action of a pushing force the components σ_{zz} and σ_{zr} of the stress tensor also arise. However, it is rather evident (and may be verified straightforwardly from the solution obtained below) that for $K \ll 1$ the stress tensor component σ_{zr} is small in comparison with the other components. Therefore, the corresponding term may be omitted from the radial equilibrium equation in (13.13), so, as it was before, $\sigma_{rr} \approx \sigma_{\varphi\varphi}$ (but now they are z-dependent). Since the channel is perfectly rigid, the radial deformation of the plug remains fixed and equal to that in (13.14): $u_{rr} = -p(1-\sigma)/E$. However, we have from Hooke's law (13.6) that

$$\begin{aligned} u_{rr} &= E^{-1}[(1+\sigma)\sigma_{rr} - \sigma(\sigma_{rr} + \sigma_{\varphi\varphi} + \sigma_{zz})] \\ &= E^{-1}[(1-\sigma)\sigma_{rr} - \sigma\sigma_{zz}]. \end{aligned}$$

So this gives us the relation between σ_{rr} and σ_{zz}:

$$\sigma_{rr} = -p + \left(\frac{\sigma}{1-\sigma}\right)\sigma_{zz}. \tag{13.15}$$

After the integration over r of the second equation in (13.13) we obtain that $\sigma_{zz} = -\frac{1}{2}r(\partial\sigma_{zz}/\partial z)$ (the constant of integration is determined by the restriction that σ_{zr} is finite at $r = 0$). Hence by now using the boundary condition $\sigma_{zr}(a) = K\sigma_{rr}$ and equation (13.15) we obtain the following equation for σ_{zz}:

$$-\frac{a}{2}\frac{\partial\sigma_{zz}}{\partial z} = K\left(-p + \frac{\sigma}{1-\sigma}\sigma_{zz}\right).$$

Its solution with the proper boundary condition at $z = 0$: $\sigma_{zz}(0) = -\mathscr{F}/\pi a^2$, is as follows:

$$\begin{aligned} \sigma_{zz} &= \frac{p(1-\sigma)}{\sigma}\left[1 - \exp\left(-\frac{2K\sigma z}{a(1-\sigma)}\right)\right] \\ &\quad - \frac{\mathscr{F}}{\pi a^2}\exp\left(-\frac{2K\sigma z}{a(1-\sigma)}\right). \end{aligned} \tag{13.16}$$

Let us now consider the variation of σ_{zz} along the plug for different pushing forces $\mathscr{F}$. If $\mathscr{F}$ is not too large then the absolute value of σ_{zz} monotonically decreases with z from the initial value $\mathscr{F}/\pi a^2$ at $z = 0$ to zero at some point $z = z_0 < L$ (for $z > z_0$ equation (13.16) is not valid in this case and we may put $\sigma_{zz} = \sigma_{zr} = 0$ there). Physically this means that the friction force acting on this part of the plug is enough to resist the pushing force in this case. The critical moment arises when $z_0 = L$. There is no equilibrium of the plug for the pushing force exceeding this critical value, so we obtain from equation (13.16) that

$$\mathscr{F}_{\min} = \frac{p\pi a^2(1-\sigma)}{\sigma}\left[\exp\left(\frac{2K\sigma L}{a(1-\sigma)}\right) - 1\right]. \tag{13.17}$$

It is seen now that for a long plug the minimum pushing force becomes exponentially large even for $K \ll 1$. The reason indeed lies in the Poisson's ratio. If $\sigma \to 0$ it then follows from (13.17) that we come back to the simple answer in this case:

$$\mathscr{F}_{\min}(\sigma \to 0) = 2\pi a L p K.$$

Problem 96

An elastic cylinder of radius R is slit along the plane formed by the axis of the cylinder and its generatrix. This cut is then extended by an external force through an angle α, ($\alpha \ll 2\pi$), and a wedge of the same size and made from the same (but unstressed) material is put into this cavity. After this, the cut is set free, i.e. no external force is present (figure 13.4). Determine the stresses in the cylinder and its elastic energy (per unit length).

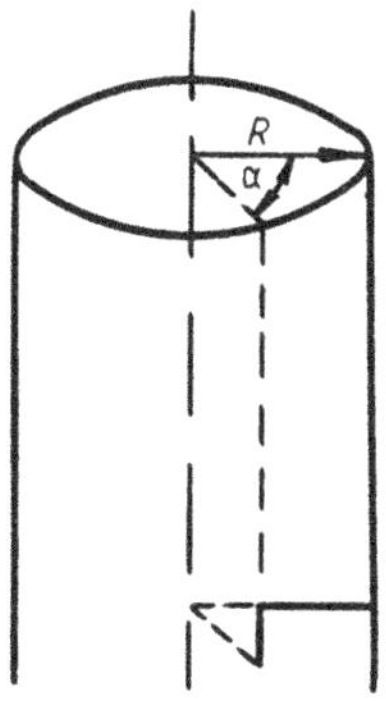

Figure 13.4 Construction of a stressed elastic cylinder with a cut.

We may now consider this cylinder as a continuous elastic body, and it follows from the geometry of the problem that the azimuthal displacement in this case is equal to $u_\varphi = \alpha r\varphi/2\pi$ (note that u_φ becomes discontinuous at the cut). In addition, the radial displacement, u_r, will also be present (the cylinder becomes inflated), and u_r depends only on r because of the symmetry: $u_r(r)$. The radial displacement has to be determined from the equation of equilibrium (13.10) for an elastic medium:

$$\nabla^2 \boldsymbol{u} + \frac{1}{1-2\sigma} \nabla \operatorname{div} \boldsymbol{u} = 0. \tag{13.18}$$

Since

$$\operatorname{div} \boldsymbol{u} = \frac{1}{r}\frac{\partial}{\partial r}(ru_r) + \frac{1}{r}\frac{\partial u_\varphi}{\partial \varphi} = \frac{1}{r}\frac{\partial}{\partial r}(ru_r) + \frac{\alpha}{2\pi}$$

and is independent of φ, the azimuthal component of (13.18) is fulfilled identically, so long as

$$(\nabla^2 \boldsymbol{u})_\varphi = \nabla^2 u_\varphi - \frac{u_\varphi}{r^2} = \frac{1}{r}\frac{\partial}{\partial r}\left(r\frac{\partial u_\varphi}{\partial r}\right) - \frac{u_\varphi}{r^2} = 0.$$

The radial component of equation (13.18) results in

$$(\nabla^2 \boldsymbol{u})_r + \frac{1}{1-2\sigma}\frac{\mathrm{d}}{\mathrm{d}r}\left(\frac{1}{r}\frac{\mathrm{d}}{\mathrm{d}r}(ru_r)\right)$$
$$= \frac{1}{r}\frac{\mathrm{d}}{\mathrm{d}r}\left(r\frac{\mathrm{d}u_r}{\mathrm{d}r}\right) - \frac{u_r}{r^2} - \frac{2}{r^2}\frac{\alpha r}{2\pi} + \frac{1}{1-2\sigma}\frac{\mathrm{d}}{\mathrm{d}r}\left(\frac{1}{r}\frac{\mathrm{d}}{\mathrm{d}r}(ru_r)\right) = 0$$

(we use here formulae in (1.1) for $\nabla^2 u$).

Hence

$$\frac{\mathrm{d}}{\mathrm{d}r}\left(\frac{1}{r}\frac{\mathrm{d}}{\mathrm{d}r}(ru_r)\right) = \frac{\alpha(1-2\sigma)}{2\pi r(1-\sigma)}.$$

By integration and taking into account that $u_r(0) = 0$, we find

$$u_r = \frac{\alpha(1-2\sigma)}{2\pi(1-\sigma)}\left(\frac{r}{2}\ln\frac{r}{R} - \frac{r}{4} + \tfrac{1}{2}Ar\right)$$

where A is a yet unknown constant, which has to be obtained from the boundary condition at the lateral surface of the cylinder, namely

$$(\sigma_{ik}n_k)|_{r=R} = 0$$

that is

$$\sigma_{rr}(R) = \sigma_{zr}(R) = \sigma_{\varphi r}(R) = 0.$$

In our case only two components of the strain tensor, u_{rr} and $u_{\varphi\varphi}$, are present:

$$u_{rr} = \frac{\mathrm{d}u_r}{\mathrm{d}r} \qquad u_{\varphi\varphi} = \frac{1}{r}\frac{\partial u_\varphi}{\partial \varphi} + \frac{u_r}{r}.$$

Thus only the diagonal components of the stress tensor differ from zero, so we need to let only one quantity tend to zero, $\sigma_{rr}(R)$. According to (13.6)

$$\sigma_{rr} = \frac{E}{(1+\sigma)}\left(u_{rr} + \frac{\sigma}{1-2\sigma}(u_{rr} + u_{\varphi\varphi})\right)$$

with

$$u_{rr} = \frac{\alpha(1-2\sigma)}{4\pi(1-\sigma)}\left(\ln\frac{r}{R} + \frac{1}{2} + A\right)$$
$$u_{\varphi\varphi} = \frac{\alpha(1-2\sigma)}{4\pi(1-\sigma)}\left(\ln\frac{r}{R} - \frac{1}{2} + A\right) + \frac{\alpha}{2\pi}.$$

After some simple calculations we find that $A = -1/2(1-2\sigma)$, and the components of the stress tensor become as follows:

$$\sigma_{rr} = \frac{\alpha E}{4\pi(1-\sigma^2)} \ln\frac{r}{R} \qquad \sigma_{\varphi\varphi} = \frac{\alpha E}{4\pi(1-\sigma^2)} \ln\frac{r}{R} + \frac{\alpha E}{4\pi(1-\sigma^2)}$$

$$\sigma_{zz} = \frac{\alpha E\sigma}{2\pi(1-\sigma^2)} \ln\frac{r}{R} + \frac{\alpha E\sigma}{4\pi(1-\sigma^2)}.$$

The density of the elastic energy is

$$\mathscr{F} = \tfrac{1}{2}\sigma_{ik}u_{ik} = \tfrac{1}{2}(\sigma_{rr}u_{rr} + \sigma_{\varphi\varphi}u_{\varphi\varphi})$$

$$= \frac{\alpha^2 E(1-2\sigma)}{16\pi^2(1-\sigma^2)(1-\sigma)}\left[\left(\ln\frac{r}{R}\right)^2 + \ln\frac{r}{R}\right] + \frac{\alpha^2 E}{32\pi^2(1-\sigma^2)}.$$

The elastic energy per unit length of the cylinder is

$$W = \int \mathscr{F}\, dS = 2\pi \int_0^R \mathscr{F} r\, dr.$$

In the integration the two logarithmic terms in $\mathscr{F}$ give the contributions to the integral, which differ only in sign, thus they compensate each other in W. So finally $W = \alpha^2 ER^2/32\pi(1-\sigma^2)$.

Problem 97

One end of a thin rod of length l and radius a, ($l \gg a$), is clamped, while another is loaded with the point weight P (figure 13.5). What is the maximum weight $P_{\max}$ that the rod can sustain, if its strength to break $\sigma_{cr} = \alpha E$ (E is the Young's modulus of the rod material, α is a numerical coefficient, and $\alpha \ll a/l \ll 1$)? The weight of the rod itself is negligible.

For $\alpha \ll a/l$ the rod will remain slightly bent even under the maximum

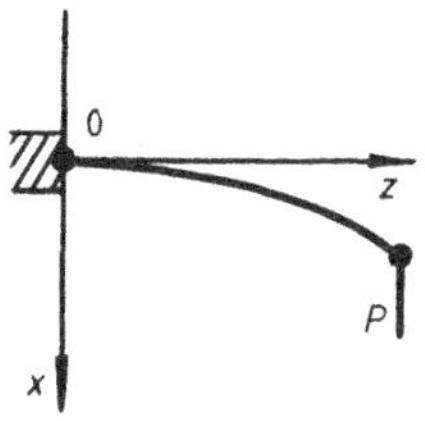

Figure 13.5. Bending of a loaded rod.

loading allowed, so its line only differs a little from the horizontal. Therefore, the equation of equilibrium for this rod may be written rather simply. Let us consider the portion of the rod located between some point with the coordinate z and the end with the weight. The moment of force created by the weight P, which is equal to $P(l-z)$, must be balanced by the moment of the internal stresses on this cross section (the bending moment). If the bending occurs in the $(x-z)$ plane, the following components of the strain tensor are present:

$$u_{zz} = x/R$$

$$u_{xx} - u_{yy} = -\sigma u_{zz} = -\sigma x/R$$

where $R(z)$ is the local radius of curvature, and x is measured from the neutral line in a cross section of the rod (in this case a neutral line is a diameter parallel to the y-axis). It then follows from (13.6) that the stress tensor only has one non-zero component $\sigma_{zz} = Eu_{zz} = Ex/R$. Therefore, a force $\sigma_{zz}\,\mathrm{d}S$ acts in the z-direction on the surface element of the cross section, so the total moment of the stress forces is equal to

$$M = \int x\sigma_{zz}\,\mathrm{d}S = \frac{E}{R}\int x^2\,\mathrm{d}S = \frac{EI}{R}$$

where $I = \pi a^4/4$ is the moment of inertia for a round rod. Thus

$$p(l-z) = EI/R(z)$$

and it is now seen that the largest curvature R^{-1}, and therefore the largest stresses arise at the clamped end of the rod, where $R^{-1}(0) = Pl/E\mathfrak{I}$. The maximum value of the stress, σ_{zz}, will be reached at the edge, where $x = a$, so that

$$\sigma_{zz}^{(\max)} = Ea/R(0) = Pal/I = 4Pl/\pi a^3.$$

Hence the maximum weight $P_{\max}$ is

$$4P_{\max}l/\pi a^3 = \alpha E$$

that is

$$P_{\max} = \alpha\pi a^3 E/4l.$$

Let us verify now the validity of this solution, which assumes that the axis of the rod changes its form slightly. Under the action of a weight this line takes the form $x(z)$, which may be determined from the equation

$$R^{-1}(z) \approx \mathrm{d}^2x/\mathrm{d}z^2 = P(l-z)/EI.$$

After integration with the boundary conditions for the clamped rod:

$x(0) = 0$, $dx/dz|_{z=0} = 0$, we have

$$x(z) = Plz^2(1 - z/3l)/2IE.$$

It is seen now that the maximum deflection, $x(l) = Pl^3/3EI$, remains really small ($x_{max} \ll l$), if $\alpha \ll a/l$.

Problem 98

Determine the critical compression force, under which a thin rod becomes unstable with respect to bending (figure 13.6).

Figure 13.6. On the bending instability of a compressed rod.

The value of this critical force, T_{cr}, depends essentially on the boundary conditions at the ends of a rod. Let us consider, for instance, a rod with hinged ends (as is shown in the figure). This means that both ends are free to turn, so the bending moment is absent there, that is the curvature of the rod is equal to zero. The unperturbed (straight) form of the rod becomes unstable when another equilibrium with a small amount of bending arises. Let the compression force be equal to T. Then the equilibrium condition for the portion of a rod between the points $(0, z)$ is as follows: the moment of the external force, equal to $-Tx$, must be balanced by the bending moment at the other end, so that

$$-Tx = EI/R(z) \approx EI\, d^2x/dz^2. \qquad (13.19)$$

The solution of this equation with the appropriate boundary conditions: $x(0) = x(l) = 0$, becomes possible firstly at $(T/IE)^{1/2}l = \pi$. Thus the critical force $T_{cr} = \pi^2 IE/l^2$. If the rod is not of round section, the moment of inertia I depends on the plane of bending. Hence, the instability threshold is determined by the amount of bending in the plane with the minimal moment of inertia.

Let us now consider another case, when both ends of the rod are clamped. Then the boundary conditions consist in turning to zero both x and dx/dz at $z = 0, l$. However, the curvature of the rod now differs from zero at the ends. Therefore, the corresponding bending moment must be included in the balance of moments. The latter is established at such a value that the required boundary condition is satisfied. Denoting this bending moment at $z = 0$ as M_0, we have now, instead of (13.19),

$$-Tx + M_0 = EI\, d^2x/dz^2.$$

The solution with $x(0) = 0$ and $dx/dz|_{z=0} = 0$ is as follows:

$$x = M_0 T^{-1}[1 - \cos(T/EI)^{1/2} z].$$

Then the condition $x(l) = 0$ gives $(T/EI)^{1/2} l = 2\pi$, so the critical force in this case becomes equal to $T_{cr} = 4\pi^2 IE/l^2$. As in the previous case, the threshold is determined by the amount of bending in a plane of minimal moment of inertia.

Chapter 14

Mechanics of Liquid Crystals

Recommended textbook:
de Gennes P G *The Physics of Liquid Crystals* (Oxford: Oxford University Press)

The state of a *nematic liquid crystal* (or *nematic*) may be defined by the *director* field $\boldsymbol{n}(\boldsymbol{r})$, where $\boldsymbol{n}$ is a unit vector representing the direction of the preferred orientation of the molecules in the neighbourhood of any point. The quantities $\boldsymbol{n}$ and $-\boldsymbol{n}$ are physically equivalent, and so only a particular axis of orientation is distinguished. In the absence of any external action the equilibrium state of a nematic corresponds to the uniform field of the directors ($\boldsymbol{n}$ = const). There are three independent deformations from this equilibrium (figure 14.1): (*a*) *splay* deformations, (*b*) *twist* deformations and (*c*) *bend* deformations.

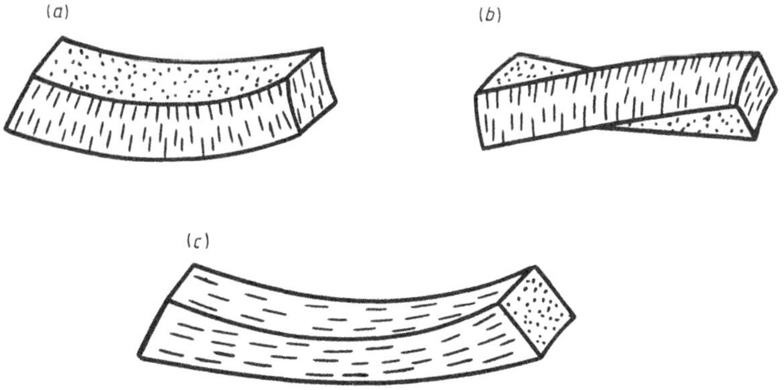

Figure 14.1. Different deformations in nematics.

Splaying is characterized by $\operatorname{div} \boldsymbol{n} \neq 0$, twisting by $\boldsymbol{n} \cdot \operatorname{curl} \boldsymbol{n} \neq 0$ and bending by $\boldsymbol{n} \times \operatorname{curl} \boldsymbol{n} \neq 0$. In a deformed state a nematic possesses an excess energy, and the density of this free energy $\mathscr{F}$ is formed by the additive contributions from all three types of deformations:

$$\mathscr{F} = K_1(\operatorname{div} \boldsymbol{n})^2 + K_2(\boldsymbol{n} \cdot \operatorname{curl} \boldsymbol{n})^2 + K_3[\boldsymbol{n} \times \operatorname{curl} \boldsymbol{n}]^2 \qquad (14.1)$$

where K_1, K_2 and K_3 are positive *elastic moduli* of the nematic: K_1 is the splay modulus, K_2 is the twist modulus and K_3 is the bend modulus.

Problem 99

Determine the equilibrium field of directors for the ground state of a *cholesteric liquid crystal* (or *cholesteric*).

Cholesterics differ from nematics in that there is no centre of inversion for their molecules. Therefore, they may also be described by pseudoscalar quantities and the elastic free-energy density $\mathscr{F}$ of cholesterics has an extra term of the form $\alpha(\boldsymbol{n}\cdot\operatorname{curl}\boldsymbol{n})$, where α is a pseudoscalar quantity. As a result their free energies $\mathscr{F}$ become as follows:

$$\mathscr{F} = K_1(\operatorname{div}\boldsymbol{n})^2 + K_2(\boldsymbol{n}\cdot\operatorname{curl}\boldsymbol{n})^2 + K_3[\boldsymbol{n}\times\operatorname{curl}\boldsymbol{n}]^2 + \alpha(\boldsymbol{n}\cdot\operatorname{curl}\boldsymbol{n}). \tag{14.2}$$

The equilibrium ground state corresponds to a minimum of $\mathscr{F}$. Unlike the nematics there will not be the uniform field of the directors, $\boldsymbol{n}$ = const. It follows from the extremum conditions for (14.2) that at the equilibrium state

$$\begin{gathered}\operatorname{div}\boldsymbol{n} = 0 \qquad \boldsymbol{n}\times\operatorname{curl}\boldsymbol{n} = 0\\ \boldsymbol{n}\cdot\operatorname{curl}\boldsymbol{n} = -\alpha/2K_2 \equiv -t_0 = \text{const.}\end{gathered} \tag{14.3}$$

This means the presence of uniform twisting in the ground state. Let us consider, for instance, the solution of equations (14.3) that depends only on one coordinate, namely z, so that $\boldsymbol{n}(\boldsymbol{r}) \equiv \boldsymbol{n}(z)$. Then the condition $\operatorname{div}\boldsymbol{n} = 0$ results in n_z = const, while the equality $[\boldsymbol{n}\times\operatorname{curl}\boldsymbol{n}] = 0$ means that $n_z = 0$. Hence we obtain the plane field of the directors (n_x, n_y), which may be represented in the form: $n_x = \cos\varphi(z)$, $n_y = \sin\varphi(z)$. Since in this case $(\operatorname{curl}\boldsymbol{n})_x = -\cos\varphi\, \mathrm{d}\varphi/\mathrm{d}z$, and $(\operatorname{curl}\boldsymbol{n})_y = -\sin\varphi\, \mathrm{d}\varphi/\mathrm{d}z$, we find from (14.3) that

$$(\boldsymbol{n}\operatorname{curl}\boldsymbol{n}) = -\,\mathrm{d}\varphi/\mathrm{d}z = -t_0$$

that is

$$\varphi = \varphi_0 + t_0 z.$$

It is now seen that the ground state of a cholesteric liquid crystal is a helicoidal structure, in which $\boldsymbol{n}$ is uniform in each of a family of parallel planes, and twists uniformly about the normal to these planes. The torsion has a full pitch of $2\pi/t_0$, but since $\boldsymbol{n}$ and $-\boldsymbol{n}$ are physically equivalent, the physical period of repetition is equal to π/t_0.

Problem 100

The domain between two parallel glass plates is filled with a nematic. The interaction between the nematic and the glass is assumed to be such that the director is constrained to lie parallel to the glass at the boundaries (see figure 14.2). Under what value of the external magnetic field $\boldsymbol{B}$ does the uniform distribution of directors become unstable (the Fredericksz transition). Consider for simplicity the moduli K_1 and K_3 to be equal ($K_1 = K_3 = K$).

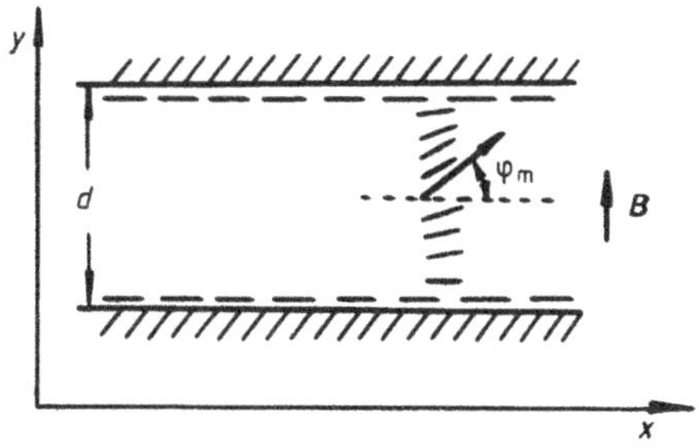

Figure 14.2. The Fredericksz transition in an external magnetic field.

The presence of an external magnetic field has an effect on the free energy of the nematic. For a sufficiently weak field this may be described by the following additional term in $\mathscr{F}$:

$$\mathscr{F} = K_1(\operatorname{div}\boldsymbol{n})^2 + K_2(\boldsymbol{n}\cdot\operatorname{curl}\boldsymbol{n})^2 + K_3(\boldsymbol{n}\times\operatorname{curl}\boldsymbol{n})^2 - \beta(\boldsymbol{n}\cdot\boldsymbol{B})^2. \tag{14.4}$$

This is the first non-zero term in the expansion in powers of the magnetic induction. In the absence of a magnetic field the directors are oriented uniformly between the plates: $\boldsymbol{n} = \text{const}$. However, if the constant β in (14.4) is positive then under the external field the free energy will be reduced while $\boldsymbol{n}$ is rotated so that it is parallel to $\boldsymbol{B}$. Hence a competition between the orientational action of the glass and that of the magnetic field arises. The result is as follows. If the magnetic field induction is less than some critical value B_{cr}, the orientation of the directors remains uniform (and parallel to the glass surfaces) throughout the entire volume. However, when B exceeds B_{cr}, such a uniform field of the directors becomes unstable, and the deformation of the nematic occurs in the stable equilibrium state.

To determine the value of B_{cr} let us derive the equilibrium equations for the director field $\boldsymbol{n}(\boldsymbol{r})$. In this particular case this field is plane (only n_x and n_y, which depend on the y-coordinate, are present). Therefore, we may introduce the angle φ by $n_x = \cos\varphi$ and $n_y = \sin\varphi$. The twist deformation is absent ($\boldsymbol{n}\cdot\operatorname{curl}\boldsymbol{n} = 0$), so the total energy W, according to (14.4), takes the form:

$$W = \int_0^d \mathscr{F}\,\mathrm{d}y = \int_0^d \mathrm{d}y\left[K_1\cos^2\varphi\left(\frac{\mathrm{d}\varphi}{\mathrm{d}y}\right)^2 + K_3\sin^2\varphi\left(\frac{\mathrm{d}\varphi}{\mathrm{d}y}\right)^2 - \beta B^2\sin^2\varphi\right]$$

$$= \int_0^d \mathrm{d}y\left[K\left(\frac{\mathrm{d}\varphi}{\mathrm{d}y}\right)^2 - \beta B^2\sin^2\varphi\right].$$

The equilibrium state corresponds to the extremum of W as a functional of $\varphi(y)$. The variation of W is

$$\delta W = \int_0^d \mathrm{d}y\left(2K\frac{\mathrm{d}\varphi}{\mathrm{d}y}\frac{\mathrm{d}\delta\varphi}{\mathrm{d}y} - 2\beta B^2\sin\varphi\cos\varphi\,\delta\varphi\right)$$

$$= \int_0^d \mathrm{d}y\left(-2K\frac{\mathrm{d}^2\varphi}{\mathrm{d}y^2} - 2\beta B^2\sin\varphi\cos\varphi\right)\delta\varphi$$

(we have integrated the first term by parts and taken into account the boundary condition $\delta\varphi = 0$ at $y = 0, d$, where the direction of $\boldsymbol{n}$ is fixed). From the condition that $\delta W = 0$ for an arbitrary $\delta\varphi(y)$ we obtain the equation of equilibrium:

$$\frac{\mathrm{d}^2\varphi}{\mathrm{d}y^2} + \frac{\beta B^2}{K}\sin\varphi\cos\varphi = 0. \tag{14.5}$$

It is seen that under any value of B the equation (14.5) has the solution $\varphi = 0$, which corresponds to a uniform field of directors. However, this solution becomes unstable at fields such as $B = B_{cr}$, when another solution of this equation arises, which has a small deformation of the nematic ($\varphi \ll 1$). Thus, equation (14.5) may be simplified as

$$\frac{\mathrm{d}^2\varphi}{\mathrm{d}y^2} + \frac{\beta B^2}{K}\varphi = 0.$$

Its solution with the boundary condition $\varphi(0) = 0$ is $\varphi \propto \sin(\beta B^2/K)^{1/2}y$, so the second boundary condition is $\varphi(d) = 0$, which gives us the critical magnetic induction:

$$(\beta B^2/K)^{1/2}d = \pi$$

that is

$$B_{cr} = (\pi/d)(K/\beta)^{1/2}.$$

Problem 101

Determine the equilibrium field of the directors in the case of the previous problem, when the external magnetic field is slightly above the critical one: $(B - B_{cr}) \ll B_{cr}$.

When the critical field is exceeded slightly, the deformation of the nematic will be weak so it is possible to retain only two terms in the expansion in powers of φ in equation (14.5):

$$\frac{d^2\varphi}{dy^2} \approx -\frac{\beta B^2}{K}(\varphi - \tfrac{2}{3}\varphi^3).$$

The first integral of this equation (which can be easily found with the help of a mechanical analogy, as in problem 92) is as follows:

$$\frac{1}{2}\left(\frac{d\varphi}{dy}\right)^2 + \frac{\beta B^2}{2K}\left(\varphi^2 - \frac{\varphi^4}{3}\right) = \text{const} = \frac{\beta B^2}{2K}\left(\varphi_m^2 - \frac{\varphi_m^4}{3}\right)$$

where φ_m is the maximum angle at which director is turned by the magnetic field (this corresponds to the point $y = d/2$, see figure 14.2). One more integration under the condition $\varphi(0) = 0$ gives the dependence $\varphi(y)$ in the following form:

$$B(\beta/K)^{1/2}y = \int_0^{\varphi} d\xi\,(\varphi_m^2 - \xi^2 - \tfrac{1}{3}\varphi_m^4 + \tfrac{1}{3}\xi^4)^{-1/2}.$$

Then the maximum angle φ_m may be determined from the following relation:

$$B\left(\frac{\beta}{K}\right)^{1/2}\frac{d}{2} = \int_0^{\varphi_m} d\xi\,(\varphi_m^2 - \xi^2 + \tfrac{1}{3}\varphi_m^4 + \tfrac{1}{3}\xi^4)^{-1/2}$$

$$= \int_0^1 dx\,[1 - x^2 - \tfrac{1}{3}\varphi_m^2(1 - x^4)]^{-1/2}.$$

This integral may be calculated approximately, when $\varphi_m \ll 1$. Then

$$\int_0^1 dx\,[1 - x^2 - \tfrac{1}{3}\varphi_m^2(1 - x^4)]^{-1/2} \approx \int_0^1 \frac{dx}{(1 - x^2)^{1/2}}[1 + \tfrac{1}{6}\varphi_m^2(1 + x^2)]$$

$$= \tfrac{1}{2}\pi + \tfrac{1}{8}\pi\varphi_m^2.$$

Thus, we easily obtain now that $\varphi_m = 2[(B/B_{cr}) - 1]^{1/2}$, so when the strength of the magnetic field is slightly above the critical value, the deformation of the field of directors is proportional to the square root of the relative excess ($B - B_{cr}$ is the excess i.e. the magnetic field above the critical field; $(B - B_{cr})/B_{cr}$ is the relative excess).

Problem 102

An external magnetic field applied to a cholesteric in a perpendicular direction to the twist axis has the effect of unwinding the structure. Determine the critical field above which the cholesteric has nematic ordering.

If the twist axis is directed along the z-axis, we can take the director in the form

$$n_x(z) = \cos\varphi(z) \qquad n_y(z) = \sin\varphi(z) \qquad n_z = 0$$

so under the external magnetic field $\boldsymbol{B} = (0, B, 0)$ the elastic energy density of a cholesteric becomes, according to (14.2) and (14.4),

$$\mathscr{F} = K_2\left(\frac{\mathrm{d}\varphi}{\mathrm{d}z} - t_0\right)^2 - \beta B^2 \sin^2\varphi \qquad t_0 = \alpha/2K_2.$$

It is seen from this expression that the magnetic field tries to orient the director along the x-axis (we assume here that β is positive), hence it really has an unwinding effect. Let us now derive the equation of equilibrium for $\varphi(z)$ from the extremum condition of the total elastic energy,

$$W = \int \mathscr{F}\,\mathrm{d}z = \int \mathrm{d}z\left[K_2\left(\frac{\mathrm{d}\varphi}{\mathrm{d}z} - t_0\right)^2 - \beta B^2 \sin^2\varphi\right]. \qquad (14.6)$$

After some simple calculations the condition $\delta W = 0$ gives the following equation of equilibrium:

$$\frac{\mathrm{d}^2\varphi}{\mathrm{d}z^2} + \frac{\beta B^2}{K_2}\sin\varphi\cos\varphi = 0.$$

The first integral of this equation is

$$\gamma^2\left(\frac{\mathrm{d}\varphi}{\mathrm{d}z}\right)^2 + \sin^2\varphi = \mathrm{const} = \kappa^{-2} \qquad (14.7)$$

where $\gamma \equiv (K_2/\beta B^2)^{1/2}$, and the integration constant, κ^{-2}, must be greater than unity, since φ is a periodic function of space. Then the dependence of φ on z is given implicitly by the elliptic integral of the first kind†

† All the information about elliptic integrals used in this problem may be found, for instance, in Abramowitz M and Stegun I A *Handbook of Mathematical Functions* (New York: Dover) §17.

$$\frac{z}{\gamma\kappa} = \int_0^{\varphi} d\xi (1 - \kappa^2 \sin^2 \xi)^{-1/2} \equiv \mathscr{F}(\kappa, \varphi).$$

The spatial period of the structure, corresponding to a variation of φ by 2π is given by

$$L = 4\gamma\kappa K(\kappa) \tag{14.8}$$

where

$$K(\kappa) = \int_0^{\pi/2} d\xi\, (1 - \kappa^2 \sin^2 \xi)^{-1/2} = \mathscr{F}(\kappa, \tfrac{1}{2}\pi)$$

is the complete elliptic integral of the first kind. Hence the elastic energy per unit volume averaged over the period is, according to (14.6),

$$\tilde{W} = \frac{1}{L}\int_0^L dz \left[K_2 \left(\frac{d\varphi}{dz} - t_0\right)^2 - \beta B^2 \sin^2 \varphi \right]$$

$$= \frac{4\beta B^2}{L} \int_0^{\pi/2} \frac{d\varphi}{d\varphi/dz} \left[\gamma^2 \left(\frac{d\varphi}{dz} - t_0\right)^2 - \sin^2 \varphi \right]. \tag{14.9}$$

The value of κ has to be determined by minimizing this free energy. By substitution of $d\varphi/dz$ from (14.7), the expression (14.9) for $\tilde{W}$ can be put into the form

$$\tilde{W} = \beta B^2 \left(2\kappa^{-2} \frac{E(\kappa)}{K(\kappa)} - \frac{\gamma\pi t_0}{\kappa K(\kappa)} - \kappa^{-2} + \gamma^2 t_0^2 \right) \tag{14.10}$$

where

$$E(\kappa) \equiv \int_0^{\pi/2} d\xi\, (1 - \kappa^2 \sin^2 \xi)^{1/2}$$

is the complete elliptic integral of the second kind, and the identity

$$\int_0^{\pi/2} \frac{\sin^2 \xi\, d\xi}{(1 - \kappa^2 \sin^2 \xi)^{1/2}} = \frac{K(\kappa) - E(\kappa)}{\kappa^2}$$

has been used. Taking into account the following differentiation rules for complete elliptic integrals:

$$\frac{dK}{d\kappa} = \left(\frac{E}{\kappa(1 - \kappa^2)} - \frac{K}{\kappa} \right) \qquad \frac{dE}{d\kappa} = \frac{E - K}{\kappa}$$

calculating $d\tilde{W}/d\kappa$ from (14.10) and equating it to zero, gives

$$E(\kappa)/\kappa = \tfrac{1}{2}\pi\gamma t_0 \qquad (14.11)$$

which determines κ as a function of γ (and, hence, the magnetic field B). Using this result, the equation for the pitch (14.8) can be written in the form

$$L/L_0 = (4/\pi^2)K(\kappa)E(\kappa) \qquad (14.12)$$

where $L_0 = 2\pi/t_0$ is the pitch in zero field. As follows from (14.11) and (14.12), as the field is increased the cholesteric unwinds and the pitch grows. For small fields, when $(\gamma t_0)^{-1} = Bt_0(\beta/K_2)^{1/2} \ll 1$, κ is small: $\kappa \approx (\gamma t_0)^{-1}$, and from (14.12) we find

$$L/L_0 \approx 1 + \tfrac{1}{32}(\gamma t_0)^{-4}$$

so the pitch grows slowly with the fourth power of the field. The transformation from a cholesteric to a nematic structure occurs at the critical magnetic field B_{cr}, when the pitch becomes infinite. This corresponds to the value $\kappa = 1$ in equation (14.11), therefore

$$\gamma_{cr}t_0 = 2/\pi$$

that is

$$B_{cr} = (\pi t_0/2)(K_2/\beta)^{1/2}.$$

Above this critical field the liquid crystal has nematic ordering.

Problem 103

Determine the distribution of the directors $\boldsymbol{n}(\boldsymbol{r})$ for a straight *disclination* with a given *Frank index m*. Consider for simplicity the moduli K_1 and K_3 to be equal: $K_1 = K_3 = K$.

Let the disclination line be directed along the z-axis. Then the distribution of directors will be plane so that the non-zero components of $\boldsymbol{n}$, n_x and n_y, depend only on the transverse coordinates, x and y. Hence we can put $n_x = \cos\varphi(x, y)$, $n_y = \sin\varphi(x, y)$. It is evident that for a given $\boldsymbol{n}$ the value of the angle φ is not specified uniquely: any quantity of the form $l\pi$, where l is a positive or negative integer may be added to φ. The presence of disclination leads to the fact, that while passing along any closed contour in the x–y plane around the origin of the coordinate system (which is the projection of the disclination line), the angle φ increases by the same value $m\pi$, where m is a positive or negative integer, called the Frank index: $\oint \nabla\varphi\, d\boldsymbol{S} = m\pi$.

The equilibrium equation for the function $\varphi(x, y)$ can be derived in the same way as has been done in problem 100. In this plane case, twisting is absent and for $K_1 = K_3 = K$ the free energy density of a nematic takes the simple form $\mathscr{F} = K(\nabla\varphi)^2$. By variation of the total elastic energy (per unit length of the disclination)

$$W = K \int (\nabla\varphi)^2 \, dx \, dy$$

the following equation of equilibrium can easily be obtained: $\nabla^2\varphi = 0$. Introducing the polar coordinates (r, χ) in the x–y plane we can conclude that the solution of interest for φ must be independent of r (otherwise the increment $\oint \nabla\varphi \, d\mathbf{S}$ would be different for circles of different radii). Therefore, we have the following equation for $\varphi(\chi)$: $d^2\varphi/d\chi^2 = 0$, that is $\varphi = A\chi + \varphi_0$. The constant A relates to the Frank index by the condition that

$$\oint \nabla\varphi \, d\mathbf{S} = m\pi$$

hence

$$\varphi = \tfrac{1}{2}m\chi + \varphi_0. \tag{14.13}$$

The resulting distribution of directors may be represented visually by plotting the 'streamlines' of the director, which are tangential to $\boldsymbol{n}$ at each point. The differential equation of these lines is (see figure 14.3):

$$\frac{1}{r}\frac{dr}{d\chi} = \cot\alpha = \cot[(\tfrac{1}{2}m - 1)\chi + \varphi_0]$$

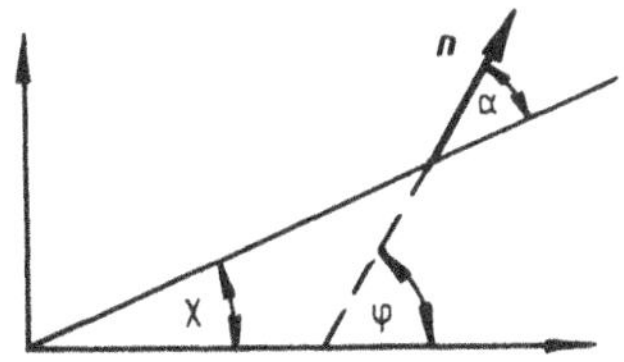

Figure 14.3. On the derivation of directors for the straight disclination.

If the Frank index of the disclination is such that $m \neq 2$, the constant φ_0 can be eliminated by the appropriate choice of the line with $\chi = 0$. The corresponding distributions of directors for the cases $m = 1$ and $m = -1$ are shown in figure 14.4. When $m = 2$, the picture depends on φ_0. Some of the other possibilities are also shown.

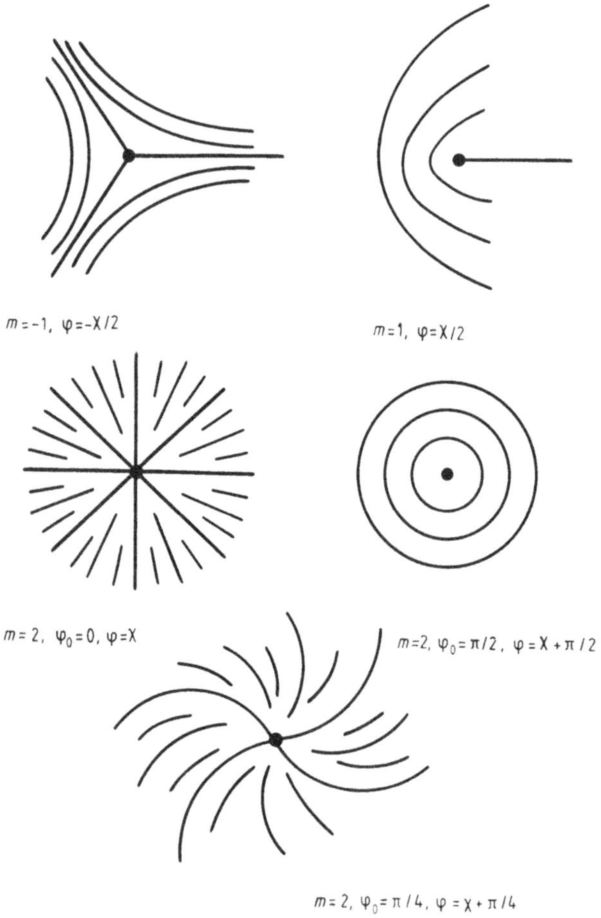

Figure 14.4. Disclinations with different Frank indexes.

Problem 104

Two parallel straight disclinations with the Frank indexes $m_1 = m_2 = 1$ are present in a nematic liquid crystal with equal moduli ($K_1 = K_3$). Plot the resulting distribution of directors and determine the force of interaction between these disclinations (per unit length).

Let the distance between disclination lines be d. Then at distances much larger than d the distribution of directors will be approximately the same as for one disclination with the Frank index equal to the sum of indexes of

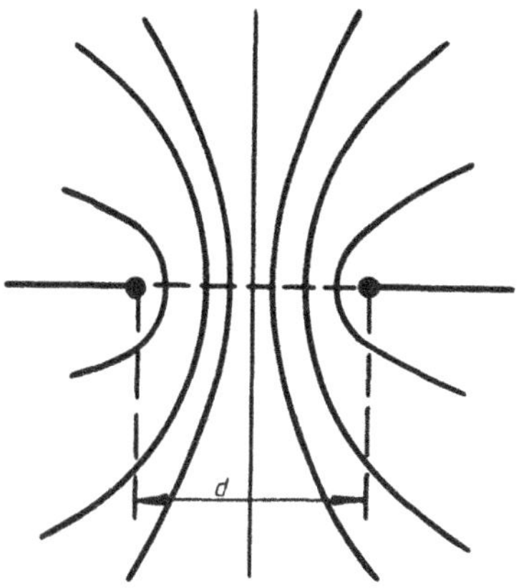

Figure 14.5. Interaction of two straight disclinations.

both disclinations. In this particular case $m_1 + m_2 = 2$, hence the resulting picture depends on the relative orientation of the disclinations. One possible case is shown in figure 14.5.

The interaction force between disclinations can be derived from considerations of energy. Let us determine firstly the elastic energy of a nematic with one straight disclination of Frank index m. This energy (per unit length of the disclination line) is

$$W = K\int (\nabla\varphi)^2\,\mathrm{d}x\,\mathrm{d}y = K\int\left(\frac{1}{r}\frac{\mathrm{d}\varphi}{\mathrm{d}\chi}\right)^2 2\pi r\,\mathrm{d}r = \frac{2\pi K m^2}{4}\int\frac{\mathrm{d}r}{r} \tag{14.14}$$

(the expression (14.13) for $\varphi(\boldsymbol{r})$ is used here).

It is seen that the resulting integral is logarithmically divergent both at small and large distances from the disclination line. Therefore, the global size of the domain occupied by a liquid crystal, L, must be used as the upper limit of the integration in (14.14). As to the lower limit, it must be of the order of the molecular size r_0, since for $r \lesssim r_0$ the deformation of a nematic is not weak, and the theory based on expression (14.1) for the free energy fails. Hence

$$W \approx \tfrac{1}{2}\pi K m^2 \ln L/r_0.$$

For a macroscopic sample of a liquid crystal the ratio $L/r_0 \gg 1$, so the energy of the disclination is proportional to the large logarithm, which has a twofold consequence. Firstly, it makes the disclination energy practically insensitive to the particular values of L and r_0, and, secondly, this energy greatly exceeds the elastic energy concentrated at the disclination core (at $r \lesssim r_0$). Namely this feature permits us to consider a disclination as a macroscopic object.

It is easy to see from the expression (14.14) that from the energy point of view such a disclination is equivalent to a straight uniformly charged line.

If its linear charge density is κ, the energy of the electric field per unit length is

$$W_{\mathrm{el}} = \int \frac{E}{8\pi} 2\pi r \, \mathrm{d}r = \kappa^2 \int \frac{\mathrm{d}r}{r}.$$

Thus it becomes clear by comparing this expression with (14.14) that the quantity $\tilde{\kappa} = m(\pi K/2)^{1/2}$ corresponds to κ. It then follows at once that the interaction of two parallel straight disclinations is equivalent to the electric interaction of two charged straight lines, so the force per unit length is

$$T = \tilde{\kappa}_1 (2\tilde{\kappa}_2/d) = \pi K m_1 m_2/d.$$

For $m_1 = m_2 = 1$ disclinations repel each other with the force $T = \pi K/d$.

This analogy can be proved formally by considering two conjugate harmonic functions,

$$\varphi_1(r, \chi) = \tfrac{1}{2}m\chi + \varphi_0 \qquad \text{and} \qquad \varphi_2(r, \chi) = \tfrac{1}{2}m \ln r$$

where the first corresponds to the distribution of directors for a straight disclination, while the second gives the electric field potential of a straight charged line.

Problem 105

Find the force acting on a straight disclination with the Frank index m, located at the distance h from the boundary plane of the nematic, where the orientation of directors is fixed (see figure 14.6).

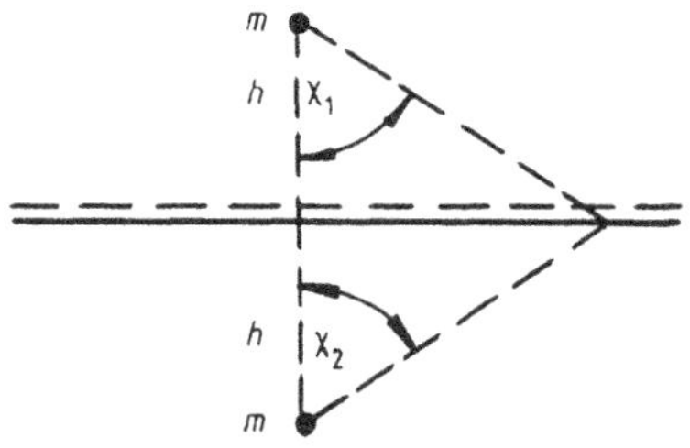

Figure 14.6. Repulsion of disclinations from a boundary plane.

This problem may be solved by exploring the electrostatic analogy mentioned above. It becomes clear from figure 14.6 that the equilibrium distribution of directors, which satisfies the boundary condition, can be formed by two disclinations: the given one and the 'reflected' disclination with the same Frank index. Indeed, the angles χ_1 and χ_2 are equal, and while moving along the boundary plane a clockwise rotation takes place for one disclination, and a counterclockwise rotation for the other. Therefore, the orientation of directors at the plane remains fixed. It immediately follows from the previous problem that such a disclination will be repelled from the boundary with a force per unit length equal to $\pi K m^2/2h$.

Problem 106

A nematic liquid crystal with equal moduli ($K_1 = K_3 = K$) occupies the domain inside a cylindrical pipe of radius R. On the boundary the directors are oriented parallel to the boundary and lie in the plane perpendicular to the axis of the cylinder. Determine the equilibrium distribution of directors in this case.

It follows from the definition of the problem that the distribution of directors will be plane with $n_x(x, y)$, $n_y(x, y)$ and $n_z = 0$ (the z-axis is directed along the axis of the cylinder). Since on the boundary circle (at $r = R$) the directors are parallel to the circle, the net rotation angle of the director while moving along this contour will be 2π. This means that disclinations must be present inside the circle, with the sum of the Frank indexes equal to two. One possible solution may be as follows: one disclination with Frank index $m = 2$ and circular 'streamlines' is located at the centre. However, this solution is unstable since it possesses excess elastic energy. This conclusion becomes evident from the following considerations.

Suppose that two very closely neighbouring disclinations with $m_1 = m_2 = 1$ appear at the centre instead of one disclination having $m = 2$. Then they will repel each other and try to increase the distance between them. However, at the same time they cannot become very close to the boundary since a boundary with a fixed orientation of directors also repels disclinations (see

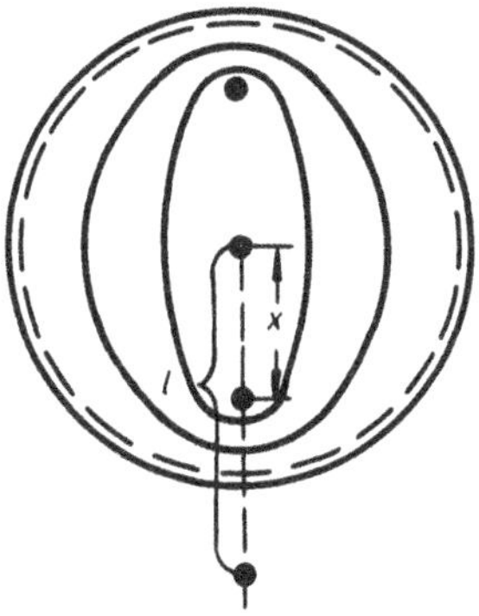

Figure 14.7. Equilibrium distribution of directors for two disclinations with $m = 1$.

the previous problem). Thus in the equilibrium state two disclinations with Frank indexes $m_1 = m_2 = 1$ will be located at such a distance x from the centre that the net force acting on each of them will be equal to zero. The corresponding solution may be found again by the reflection method as in similar electrostatic problems. This means that the prescribed boundary condition for the directors will be satisfied, if to each disclination located inside the cylinder at a distance x from the centre, there corresponds a 'reflected' disclination with the same Frank index, but located outside the cylinder at a distance $l = R^2/x$ from the centre. It then follows from the balance of forces that

$$\frac{1}{l - x} = \frac{1}{2x} + \frac{1}{l + x}$$

so $x = R/\sqrt[4]{5}$. The corresponding field of 'streamlines' is shown qualitatively in figure 14.7.

Index